The Politics of Crisis Management in China

The Politics of Crisis Management in China

The Sichuan Earthquake

Sonny Shiu-Hing Lo

LEXINGTON BOOKS
Lanham • Boulder • New York • London

Published by Lexington Books
An imprint of The Rowman & Littlefield Publishing Group, Inc.
4501 Forbes Boulevard, Suite 200, Lanham, Maryland 20706
www.rowman.com

16 Carlisle Street, London W1D 3BT, United Kingdom

British Library Cataloguing in Publication Information Available

Library of Congress Cataloging-in-Publication Data

Lo, Shiu Hing, 1963-
The politics of crisis management in China : the Sichuan Earthquake / Sonny Shiu-Hing Lo.
pages cm
Includes bibliographical references and index.
ISBN 978-0-7391-3952-3 (cloth : alk. paper) -- ISBN 978-0-7391-3954-7 (electronic)
1. Wenchuan Earthquake, China, 2008. 2. Crisis management--China. 3. Emergency management--China. 4. Earthquake relief--China--Sichuan Sheng. 5. China--Politics and government--2002- I. Title.
HV6002008.W46 L6 2015
363.34'958095138090511--dc23
2014034966

∞ ™ The paper used in this publication meets the minimum requirements of American National Standard for Information Sciences Permanence of Paper for Printed Library Materials, ANSI/NISO Z39.48-1992.

Printed in the United States of America

Contents

List of Tables

Acknowledgment

This book could not have been completed without the support of A. Lenore Lautigar, the former editor of Rowan and Littlefield, who accepted my book proposal in 2012 and gave me a golden opportunity to begin this book project. I have to extend my gratitude to editor Justin Race who kindly granted me an extension at a time after I relocated from Canada to Hong Kong. The extension period allowed me to complete the entire manuscript. Finally, I have to thank my two former research assistants, Olivia Poon and Iris Chan, for their help in collecting the relevant literature for my book project, and to Minxing Zhao who provided me with some Chinese literature on how mainland Chinese scholars have viewed the progress of post-earthquake relief and reconstruction work in China. Indeed, I alone shoulder the responsibility of any error in this book.

Abbreviations

ATV	Asia Television (Hong Kong)
CCH	Chief Command Headquarters
CCP	Chinese Communist Party
CCTV	China Central Television
NIDS	National Institute for Defense Studies (Japan)
PAP	People's Armed Police
PLA	People's Liberation Army
ROC	Republic of China (Taiwan)
SARS	Severe Acute Respiratory Syndrome
SDF	Self-Defence Force (Japan)
UN	United Nations

Source: National Administration of Surveying, Mapping and Geoinformation of the People's Republic of China, http://219.238.166.215/mcp/index.asp, access date: June 28, 2014.

Introduction

Crisis management in dealing with the sudden eruption of earthquake is arguably a political process that not only tests the capability of a regime but can also plunge a government concerned into a serious crisis of legitimacy, especially if the regime or governmental responses are weak and ineffective. The context and content of governmental responses to earthquake are central to our understanding of whether crisis management in a state is swift and effective, thus maintaining and even consolidating its own legitimacy. In a sense, the occurrence of earthquake in any state presents both a crisis and an opportunity, a crisis in that any ineffectual response would delegitimize the regime further and an opportunity because rapid rescue of victims and swift recovery can arguably consolidate regime legitimacy.

Crises can be political, economic-technological, social, and natural (Kim and Lee, 2001). Political crises embrace wars, armed strife, military coups, terrorism, hijacking, and subversive activities. Economic-technological crises include architectural collapses, flammable chemical explosions, radioactive or toxical chemical spills, and environmental pollution. Social crises range from riots to the outbreak of diseases, and from violence to labor strikes. Natural crises are numerous, including flood, typhoons, hurricane, earthquakes, droughts, cold-weather damage, and storms.

This book is going to use the 2008 earthquake in Sichuan as a case study of studying how the People's Republic of China (PRC) tackled crisis management. In particular, a number of questions are crucial to our understanding of how effective the PRC state was in dealing with disaster outbreak. First, whether the government's responses were sufficient in rescuing the victims in the first 48 hours. Second, what were the responses from the state, citizens, and groups to the earthquake? Third, how was the People's Liberation Army (PLA) mobilized to tackle the sudden eruption of the earthquake?

Fourth, what was the eventual reconstruction work that aimed at achieving recovery of those towns and cities destroyed by the powerful earthquake? Finally, were the PRC responses comparable to other similar natural disasters from a comparative perspective? This book aims at answering these crucial questions shaping the legitimacy of the Chinese state in response to the Sichuan earthquake in 2008.

EARTHQUAKE CRISIS MANAGEMENT AND REGIME LEGITIMACY

The PRC state had been challenged by the magnitude 7.8 Tangshan earthquake on July 28, 1976, when almost a quarter of a million people were crushed to death in the counties surrounding the Qinglong county, which according to Jeanne-Marie Col and Jean Chu had actually set up an earthquake office, organized procedures for proper responses, trained government officials, and had activated the earthquake plan to deal with citizens' safety (Col and Chu, 2001: 581). Two weeks prior to the earthquake, Qinglong officials detected abnormal signals for the Beijing-Tianjin-Bohai-Zhangjiakou region that pointed to a possible earthquake. On July 24, the committee of the Chinese Communist Party (CCP) in Qinglong county listened to a report presented by an official of the science committee that detected possible earthquake eruption. The party committee decided to strengthen leadership at all levels to monitor the situation, to promote the dissemination of information to citizens if any earthquake broke out, to improve communications and reports, and to place earthquake equipment in air-raid resistant shelters (Col and Chu, 2001: 583). On July 25 and 26, each community in the Qinglong county held an emergency meeting to teach villagers on how to cope with any possible disaster. County and village broadcasts were made to instruct villagers not to close the windows and doors at night so that they would be able to leave their homes quickly in the event of feeling any earthquake movement. When the earthquake struck Tangshan on July 28 at 3:42 AM, over 180,000 buildings in the county collapsed, but interestingly only one person died of heart attack, while the city of Tangshan and all its surrounding counties witnessed 240,000 deaths and 600,000 injuries (Col and Chu, 2001: 584). Col and Chu concluded that the preparation in Qinglong county actually saved many lives—a situation that was also identified and praised by the United Nations, which regarded the Qinglong county's responses as an example of the "best practice in public administration (Col and Chu, 2001: 584)."

According to Col and Chu, the Qinglong's responses were marked by several characteristics. First and foremost, public administrators in the State Council took various initiatives, including the formulation of Document

Number 69, which told seven municipalities and provinces in Northeast China that a major earthquake would be likely and which urged them to take precautionary measures. The document explicitly warned that the earthquake magnitude would amount to something between M6 and M8. Apart from the use of governmental documents, public administrators in the Qinglong county also ensured the implementation of taking precautionary measures, collected the necessary geological data to detect and interpret any earthquake occurrence, delegated the monitoring work to various earthquake detecting stations, promulgated an action plan with detailed instructions on how other officials and citizens should react to any earthquake, and finally developed a shared perception of the risks of coping with the sudden earthquake.

At the level of citizens, the Qinglong county was punctuated by an increase in community awareness of the possible earthquake through widespread and intensive public campaigns, including the phenomena of citizens receiving pamphlets, watching movies, participating in drills, and discussions about earthquakes. Employees in factories and mines were also mobilized and trained to monitor earthquake precursors by using instrument readings and observational data (Col and Chu, 2001: 587). On July 24, local residents were mobilized to move furniture from schools indoors to open spaces while holding classes outdoors; merchants were encouraged to construct shelving outdoors under plastic sheets; other residents even built makeshift tents and camped outdoors; people practiced leaving their homes and offices; citizens assemblies were held to educate the public on earthquake preparedness strategies; and members of the public patrolled on the streets to supervise preparatory activities. In short, citizens were fully mobilized in anticipation of the earthquake.

With regard to the role of earth scientists, two weeks prior to the earthquake, they had actually shared geological information and discussed geoelectric, seismic, meteorological, and hydrologic data at a conference held in Tangshan. In particular, they paid attention on the crustal stress over time. As early as 1972, the Seismo-Geological Brigade of the State Seismological Bureau had established a network of some 100 professional crustal stress-monitoring stations in which 24 were located in North China (Col and Chu, 2001: 589). Although this crustal stress identification strategy was successful in determining the location and the timing of the 1975 Haicheng and 1976 Tangshan earthquakes in the PRC, "the lack of familiarity with some important aspects of precursory signals (e.g., amplitude vs. length of time of signal) resulted in earthquake predictions of M5 (Col and Chu, 2001: 590)." It appeared that the State Seismological Bureau's prediction of the magnitude of the Tangshan earthquake was slightly less accurate than the scientists working at the Qinglong county, for the latter anticipated an earthquake with a magnitude ranging from M6 and M8.

Col and Chu have come up with the interactions between public administrators, citizens, and scientists on the one hand and their level of participation in mitigating the impacts of any earthquake on the other hand. They made a number of following observations from the Qinglong county's experiences in tackling the possible earthquake, which eventually took place in July 1976.

> The Qinglong county experience shows us that tragedies even from great earthquake disasters can be successfully prepared for. The main ingredient is awareness that (1) large earthquakes are preceded by numerous precursors; (2) there are a variety of disaster mitigation methodologies available to monitor for changes in the natural environment related to earthquakes; and (3) while today's urban metropolises are physically more than ever vulnerable to the devastation of natural events, it is through the networking of all levels of society and the open communication of disaster information that these mega-communities will achieve earthquake early warning successes (Col and Chu, 2001: 594).

Table 0.1 illustrates how they perceive the natural disaster's outcomes as a result of the complex dynamic relations between participation and the roles of citizens, public administrators, and scientists. While the level of participation shapes the outcomes of natural disasters directly or indirectly, Col and Chu have neglected the possible impacts on regime legitimacy. Arguably, if the involvement of members of the public officials and scientists was minimal, the impacts on regime legitimacy would likely be more detrimental. In other words, the lack of crisis preparation can plunge a regime into a serious crisis of legitimacy in the event of an outbreak of the earthquake. Moreover, a moderate level of citizen, official, and scientist participation would mitigate the negative impacts of earthquake occurrence on the regime legitimacy, which is contingent upon the measures taken by the government concerned. If the government can react quickly to earthquake outbreak by taking relief measures swiftly, rescuing citizens effectively, and implementing reconstruction work rapidly, regime legitimacy can be protected. Conversely, sluggish governmental responses to earthquake would de-legitimize the regime. Finally, if the level of participation on the part of citizens, officials, and scientists is high, then the impacts on regime legitimacy are arguably positive. The effective warning systems introduced by the government and its related communication strategies and preparedness would even consolidate the regime legitimacy. This book will adopt this modified framework, which originates from Col and Chu's contribution, to assess how the PRC government at various levels responded to the 2008 earthquake in Sichuan province.

MANAGING EARTHQUAKE CRISIS FROM A COMPARATIVE PERSPECTIVE

The literature on earthquake crisis management is voluminous. Some scientists have argued that the casualties of earthquakes are strongly related to the earthquake strengths, occurrence time (day or night), and the distribution of population in the affected areas (Chen, Mi, Huang, 2005). They have found that earthquakes in the PRC from 2001 to 2003 demonstrated such relationships. Others have discovered that earthquake engineering has several new challenges, including seismology, tsunamis, geotechnical engineering, buildings, lifelines such as pipelines and bridges, risks assessment, and public policy (Morelli, 2004). In the area of buildings, for instance, it is important for scientists and government officials to explore better performance of buildings in earthquakes of different intensities, embracing issues like soil foundation and building materials that can instantly sense a change and command an appropriate building component response (Morelli, 2004: 2). In general, scientists have shown their research interests in the scientific aspects of earthquake management instead of focusing on the role of governmental responses.

Other scientists, however, have given concrete suggestions on how governments should prepare for earthquake crisis management. For example, earthquake engineering expert Nateghi-A used the case of Iran to study how its government can improve the earthquake crisis management system. Nateghi-A asserted that Iran, one of the countries in the world vulnerable to earthquakes, tended to focus on the rescue operations but it lacked a master plan of dealing with any earthquake's occurrence (Nateghi-A, 2000: 202). Other weaknesses in the disaster management system included a complex bureaucracy without clear allocation of tasks among the agencies dealing with any earthquake, the lack of communications among different governmental organizations, the reliance on the Ministry of Interior to execute the emergency operations, and the absence of an office under the leadership of either the prime minister or the president to handle any crisis situation.

Nateghi-A and Izadkhah also utilized the case of Iran to advocate the importance of developing earthquake disaster management planning in health care facilities (Nateghi-A and Izadkhah, 2004: 130). They suggested that detailed planning in dealing with health care facilities should be developed, including the need to protect the lives of health care personnel, patients and visitors inside and outside hospital facilities during the earthquake crisis, the need to evacuate them, the use of emergency facilities after the earthquake, the use of emergency power systems and emergency shelters, and the mobilization of rescue personnel after the earthquake. They conclude that "the impossibility of predicting the occurrence of earthquakes and locating of Iran in the Alpine-Himalayan seismic belt as one of the most active tectonic

Table 0.1 The Participation of Citizens, Public Administrators and Scientists and the Outcomes of Natural Disasters

Participation	*Citizens*	*Public Administrators*	*Scientists*	*Natural Disaster Outcomes*
Unaware, little or no action	Ignorant, unaware	Too busy with daily affairs to consider any natural disaster	Do not think about how natural disasters affect people	Total tragedy, serious crisis of **regime legitimacy**
Educated, Moderate	Aware, take precautionary measures	Gather information, formulate plans, assess relief needs, but lacking standards for evaluating the accuracy of predictions	Note and record their predictions, produce predictions with a degree of probability	Tragedy, massive relief, major impacts on social and economic development, **regime legitimacy** depends on governmental responses
Aware, high level of alert	Citizens educate themselves, make preparation, organize drills, and monitor the situation	Organize drills, educate the public, mobilize media to prepare for the worst case scenario together with the private sector and non-governmental groups	Share and discuss precursor signals with scientists, citizens, and officials	Early warning success, evacuations of the community, lives and properties, minimize deaths and relief costs, ensure available resources for sustainable development, and **regime legitimacy** can be protected and consolidated

Source: Col and Chu, 2001: 591. Adapted from Col and Chu's table.

regions of the world necessitate a comprehensive disaster management planning in the urban and rural areas of the country. Planning for earthquakes in health care facilities especially hospitals is of great importance" (Nateghi-A and Izadkhah, 2004: 135).

Other Iranian scholars have emphasized the need for Iran to develop better risks management plans at the schools level. Hosseini and Izadkhah emphasized the importance of developing a national-level emergency plan with regard to the three stages of the earthquake: before, during, and after the earthquake (Hosseini and Izadkhah, 2006: 649). Special attention should be paid to schools, they argued, including the need for training school personnel, developing materials of training, and educating the children on how they should cope with any earthquake. In short, educating the teachers and children on their awareness, preparedness, and responses to the outbreak of earthquake is necessary. Detailed plans would have to be formulated, embracing drills, psychological preparation, first-aid training, building safety, the formation of rescue and relief teams, fire extinguishing teams, and the identification of roles and responsibilities of headmasters, school teachers, and representative groups of the children. Disaster management councils should ideally be established under all the schools. Hosseini and Izadkhah developed a comprehensive emergency response plan for the sake of protecting Iranian schools from the disruptive impacts of any earthquake in the future.

Some Japanese social scientists have focused their study on how the Japanese government responded to earthquakes. Masaru Sakamoto studied the lessons of the Great Hanshin-Awaji earthquake that hit Kobe and its surrounding areas in Japan in 1995 as a case study to assess the inadequacies of the Japanese crisis management (Sakamoto, 2001: 559). The Kobe earthquake, which was measured 7.2 on the Richter scale and killed 6,308 citizens, de-legitimized the Japanese government, whose leaders including Prime Minister Tomiichi Murayama and the Governor of Hyogo prefecture were criticized for being "slow" in their responses (Sakamoto, 2001: 565). Japan had been attacked by many earthquakes in its history, including the 1923 Kanto earthquake with a magnitude of 7.9 and causing 99,331 deaths and 43,476 people who were missing (Sakamoto, 2001: 561). While the disaster prevention plans at the local level failed to respond to the Hanshin-Awaji earthquake effectively, the local governments responded to the emergency situation through coordination work. The Kobe earthquake marked the importance of volunteering activities on the part of 1.3 million citizens, who helped the victims and themselves during the emergency.

Nevertheless, Sakamoto criticized the Japanese government prior to the 1995 earthquake as having defects in its security system, especially the lack of legal provisions relating to the establishment of a command center that could mobilize the Self-Defense Forces and the police (Sakamoto, 2001: 568). Moreover, the prime minister before 1995 did not have the power to give directives to ministers in case of the outbreak of emergencies. The government information system, including collection, analysis, assessment, and dissemination of emergency data, was inadequate in response to any earthquake occurrence. Intergovernmental relations, according to Sakamoto, were complex and bureaucratic, thus hampering any swift response to the earthquake in 1995.

After the Kobe earthquake, the Japanese government undertook administrative reform to empower the prime minister to take direct control of emergencies by strengthening the role of the Cabinet Security Affairs Office. It also revised the Basic Law on Disaster Countermeasures and increased the role and ability of the prime minister's office to gather information on natural disasters and take immediate measures to deal with any crisis situation. Furthermore, the Anti-Disaster Basic Law was revised in December 1995 to allow municipal government leaders to request the dispatch of Self-Defence Forces for any disaster relief efforts. Overall, Tokyo learned a bitter lesson from the Kobe earthquake and was determined to respond to any future crisis in a far more effective manner.

Like Sakamoto, Akira Nakamura used the Kobe earthquake to examine how the Japanese government responded to the crisis and its political implications. Nakamura stressed that the Center for Emergency Program Operation was actually established by the national government in the afternoon of the day of the Kobe earthquake, while the Cabinet later founded the Headquarters for Emergency Management that was chaired by the prime minister (Nakamura, 2000: 27). The center was led by a minister who was also sent to Kobe to lead government officials at the national level in the rescue operation, including officials from the land, police, Self-Defence Forces, finance, education, health, welfare, home affairs, and transportation. Officials from the national, prefectural, and city levels met frequently to tackle their coordination and rescue efforts. Unlike Sakamoto, Nakamura argued that the Japanese governmental response was actually swift.

However, Nakamura noted that the Kobe city government officials affected by the earthquake were displeased with the way in which they were treated as observers in a formal meeting with the officials from the national government and Hyogo prefecture (Nakamura, 2000: 28). Kobe officials felt that they were ignored and therefore left the conference meeting. Moreover, the governor of the Hyogo prefecture, Kaihara Toshitami, stole the media limelight while the Kobe mayor Sasayama Yukitoshi was a reserved politician relatively ignored by the mass media but quietly focused on the reconstruction work. Nakamura also made an insightful observation on the political aftermath of the Kobe earthquake. He wrote:

> Crisis management shows the same trend all over the world. Immediately after a disaster, citizens and government display keen interest in crisis control, arguing for increased budgets and personnel to deal with another disaster. However, as time passes, the popular interest in crisis management lapses and, as often as not, the issue becomes the target of cutbacks and streamlining. The

> only time that crisis management gains legitimacy is when a country is stricken by a disaster. This traditional axiom, unfortunately, appears to hold true in the contemporary scene in Japan (Nakamura, 2000: 29).

Nakamura's observation on the short-term concerns about crisis management in Japan after the 1995 Kobe earthquake was noteworthy, although he did not explicitly focus on how the Kobe earthquake impacted on the legitimacy of the Japanese government. However, it can be argued that both Sakamoto and Nakamura implied that the Kobe earthquake and the related sluggish governmental responses at the national level did undermine the regime legitimacy to a certain extent, thus necessitating remedial action and measures taken by Tokyo after the earthquake. Their common views of the impact of the Kobe earthquake on Tokyo were that it was indirect rather than direct, and that the regime's public image was undermined due to the insufficient preparedness.

Unlike Sakamoto and Nakamura, crisis management expert Robert Heath contended that the Kobe earthquake demonstrated the failure of the Japanese crisis responses. He argued that the Japanese community and organizational culture, which tended to promote collectivist rather than individual action, to develop strong tendency of avoiding uncertainty, and to engender "large power distance" between citizens and leaders, shaped the slowness of the governmental responses to the earthquake (Heath, 1995: 13). Heath went so far as to argue that the Japanese responses lacked integration, coordination, communication planning, and strategic management. Pre-earthquake exercises were few, intergovernmental interactions were of minor nature, escalation of responsive management was slow, and systematic strategic planning was apparently weak (Heath, 1995: 23). He concluded in the following way:

> Perhaps the greatest lesson to be gained from the Kobe earthquake is that response managers do have very little time in which to get the response "right." Time, information, and cost in resources become the key elements fixed in the task of a response manager . . . The best means of gaining more time and information, and of reducing the cost in resources, is to develop plans that work as guides to actions, and to promote and develop the concepts and practices of strategic preparedness in response organizations and their host communities. Had these approaches been adopted in time, Kobe might have emerged from an earthquake with fewer casualties and loss of resources, and with a greater belief within the Kobe and international communities that everything that was possible was attempted as quickly and effectively as the effects of that earthquake allowed (Heath, 1995: 24).

Some scholars have focused their study on the post-earthquake relief and reconstruction work. For instance, Antonios Pomonis has examined how the Greek government responded to the Mount Parnitha earthquake of September 7, 1999. He identified both the strengths and weaknesses of the state responses to the earthquake. Although damages to 42,000 buildings were considerable and 143 people died (Pomonis, 2002: 172), Pomonis argued that the Greek government was successful in its post-earthquake relief efforts, repairing many buildings and recovering lots of lost properties. However, low-income families would need more assistance from the government. Pomonis noted that, unlike other countries in the world, earthquakes in Greece did not bring about high human casualties because of three main factors. First, Greece is not densely populated and strong earthquakes occur in sparsely populated regions. Second, the magnitude of earthquakes is usually moderate, ranging from 6 to 6.9. Third, at least 60 percent of Greece's buildings have been made "from reinforced concrete frames with unreinforced infill masonry (without reinforced concrete shear walls)," thus having conservative designs and infill walls and bringing about "significant reduction in the vulnerability of this type of structure" to earthquakes (Pomonis, 2006: 193). In brief, the design of buildings in Greece appeared to become a mitigating factor in any earthquake's disruptive impacts.

Other social scientists have emphasized the importance of the responses from the subnational and local governments to natural disasters, including hurricanes and earthquakes. Jeanne-Marie Col compared and contrasted the local governmental responses in the case of Hurricane Katrina in New Orleans in August 2005 with that in the Tangshan earthquake in China in 1976. She focused on three of the four stages of the occurrence of natural disasters: (1) preparedness in which the state has plans, training, and exercises, (2) mitigation in which a sustained effort is made to reduce risks to people and property, and (3) response during which immediate action is taken to save lives and protect properties. Recovery was outside Col's analysis because it would take some time to study post-earthquake reconstruction. Hurricane Katrina led to the loss of 1,330 lives and the displacement of 770,000 people while 300,000 homes were destroyed (Col, 2007: 117). Col argued that while the New Orleans government failed in the three dimensions of preparedness, mitigation, and responses, the Qinglong county in China did well in all the aspects. She drew a number of lessons to be learned from the American and Chinese experiences. First, local governments must be able to act decisively and earlier in preparation for any natural disasters. Second, local governments must be supported by higher-level governments in disaster preparation and mitigation efforts. Col added that when Hurricane Katrina struck New Orleans, many local governments were not familiar with the Department of Homeland Security's National Response Plan. Third, citizens must participate in all the phases of preparation and implementation of emergency management measures—a phenomenon that could be found in Qinglong county but not in New Orleans. Finally, linking scientific information to public administrative action in response to natural disasters is of critical importance.

Col therefore concluded that New Orleans lagged behind the Qinglong county in response to the imminent natural disaster.

The emphasis on governmental responses to earthquake occurrence also includes public policy dimensions. Mary Comerio has stressed that governments must formulate and implement better public policy for the sake of reducing earthquake risks, including the availability of better insurance policies and building codes, the need for better land-use planning, the adoption of protective measures on the part of public utilities, large corporations, museums, and educational institutions, and the enforcement of mitigation plans (Comerio, 2004: 403). Although Comerio's discussions covered not merely earthquakes but also floods, tornadoes, hurricanes, and storms, her emphasis is of public policy nature, calling for public responsibility of dealing with natural disasters in a more effective way.

As with Comerio, Alpaslan Ozerdem and Sultan Barakat have used the case of the Marmara earthquake in Turkey in August 1999 to highlight the significance of effective public policy measures in response to the outbreak of earthquake. Ozerdem and Barakat have stressed the importance of sustainable development policy, which should ideally be implemented before any natural disaster to address poverty issues (Ozerdem and Barakat, 2000: 427). On the other hand, they also pinpoint to the need for sustainable development after earthquakes so that reconstruction work can target at those who need the state's assistance and rescue. Positively speaking, sustainable development can reduce vulnerability to earthquake disaster, which in turn can provide developmental opportunities. Negatively speaking, however, poor developmental policies can increase vulnerability to natural disaster, which in turn can retard development (Ozerdem and Barakat, 2000: 428). Both Ozerdem and Barakat have argued for the need of governments to adopt four mitigation strategies of coping with earthquakes: better legislation in dealing with building, land-use, and earthquake-safety regulations as well as preparedness planning; education and training of key professionals including engineers, planners, and media workers; the adoption of urban management tools of providing earthquake-resistant buildings and infrastructural lifelines (such as water, electricity, communication, and roads); and the utilization of better planning tools like environmental impact analysis, hazard impact analysis, and on-site and off-site safety plans. They conclude that the government authorities in Turkey failed to provide a good example in the construction of earthquake-resistant buildings because of the lack of control mechanisms, political favoritism, corruption, and the complacency of state institutions (Ozerdem and Barakat, 2000: 438). Moreover, the mobilization of the Turkish army in response to the earthquake disaster must be considered, while the deployment of financial, institutional, and legal assistance to civil society organizations in response to earthquake disaster would be imperative so as to create a wider societal response network. In a nutshell, a more holistic public policy approach to tackling earthquake disaster is a must.

It can be said that a majority of social scientists have tended to focus on the preparedness of governments in anticipation of, and in response to, earthquake disasters. Michael Hooper has studied the 1994 Northridge earthquake in California and found that residents had been educated about the possible occurrence of any earthquake. Citizen preparedness had been underway in the late 1980s. However, Hooper cautioned that the government authorities in California should continue to educate the citizens on earthquake emergency plan. He concluded: "Unquestionably, Southern Californians are among the most earthquake-educated collectives in the country. If so many of them 'cannot get it right' (in terms of initial safety actions), what will be result of major earthquakes afflicting parts of the country where such events are far less 'rehearsed'" (Hooper, 1999: 704).

As with Hooper, Richard Vedder uses the case of the 2005 Hurricane Katrina to highlight the significance of the need to improve disaster planning. He has argued that private companies should develop better policies, including information technology, in response to any natural disaster. Moreover, the places being affected by natural disasters should be better prepared in terms of having emergency generators, storing documents electronically, evacuating people effectively, and avoiding any further environmental contamination. The Americans, according to Vedder, should also understand their own duties and obligations during emergency situations, while enhancing communications among themselves. Finally, information technology professionals should prepare for the crisis scenario in which credit cards, debit cards, and automatic teller machines would be rendered useless for an indeterminate period of time. Vedder concludes in the following way: "Good disaster planning requires broad and flexible thinking. If your city has to be evacuated, an IT plan focused on computer security and data backup will simply not do. You will also need to coordinate with your department's response with that of the company as a whole" (Vedder, 2006:44). In short, comprehensive planning with some degree of flexibility is ideally the hallmark of an improved disaster planning.

CHAPTER ORGANIZATION

As mentioned before, this book will use the case of the 2008 Sichuan earthquake in the PRC to illuminate the responses of the Chinese government at the central, provincial, and local levels to the tragic crisis, focusing on the impacts of the responses on regime legitimacy. This book will be organized into the following way: Chapter 1 will focus on the critical 72 hours after the earthquake to see how the government responded to the disaster; chapter 2 will probe the role of interest groups, citizens and even Tibetan activists who support their self-determination after the outbreak of the earthquake; chapter 3 will examine how the People's Liberation Army (PLA) tackled the crisis; chapter 4 will explore the post-earthquake reconstruction work and policy learning from the 2008 Wenchuan earthquake; and chapter 5 will compare and contrast the PRC's crisis management from a historical perspective and with other countries, particularly the Chinese responses to the 2008 earthquake in comparison with the reactions of another socialist regime, Myanmar, to the cyclone that also took place in May 2008. The conclusion will sum up this book's main findings, arguments, and implications for the study of earthquake management.

Chapter One

The Critical 72 Hours after the Sichuan Earthquake

At 14:28 PM on May 12, 2008, an earthquake of magnitude 7.8 struck Wenchuan county in the Sichuan province of the PRC. Many surrounding cities felt the tremors of the earthquake, including the cities of Beijing, Shanghai, Taipei in the Republic of China (ROC), Hanoi in Vietnam, and Bangkok in Thailand (*Ming Pao*, May 12, 2008: A17). The earthquake's epicenter was located at 92 kilometers in the northeast of Chengdu. Although the Three Gorges Dam along the Yangtze River was not damaged, the earthquake lasted for two to three minutes and generated the most devastating impacts on the Wenchuan county. Deaths were immediately reported in the city of Chongqing, where a primary school collapsed and killed four students. The response of the central government in Beijing was swift; Primier Wen Jiabao decided to go to the affected areas and commanded the rescue operations (*Ibid*).

The response of the People's Liberation Army was swift. Some 30,000 troops would join the 20,000 soldiers who were already searching for survivors in crumpled homes, schools, and hospitals in Sichuan province (*Toronto Star*, May 14, 2008: A1). On May 13, it was reported that 9,219 people died in eight provinces and cities after the Wenchuan earthquake with 60,000 citizens who were missing (*Ming Pao*, May 13, 2008: A1). The Sichuan health department sent 24 medical teams to rush to the most affected regions including Wenchuan county and Mianyang city. Li Shiming, the commander of the Chengdu Military Area Command, said that once the weather took a turn for the better, the army would start airlifting food and medicines to Wenchuan (*Toronto Star*, May 14, 2008: A1).

THE RESPONSE OF THE GOVERNMENT ON MAY 12 AND 13

The most notable response from the central government in Beijing was the action taken by Premier Wen, who immediately led the Chief Command Headquarters (CCH) of the Earthquake Relief and Rescue Operation under the State Council. The CCH set up eight work groups, including rescue, medicine, life resettlement, and public order. On May 12, Wen flew to Sichuan province immediately together with the minster of public security, Meng Jianchu, and other officials. On the morning of May 13 at 7 AM, Wen held the second meeting with the members of the CCH, coordinating the rescue operations and ordering immediate action on the part of the central, provincial, and local governments. On the day of the earthquake attack, Wen arrived at Dujiangyan city's schools, which had collapsed and buried hundreds of students under the rubble. Wen said: "I knew the situation and came to the scene at once. I feel sad as with others because their lives are so important. As long as there is a glimpse of hope, we have to try our best to rescue people" (*Ming Pao*, May 13, 2008: A1). When Wen saw the dead bodies, he bowed to them three times and urged the rescue workers to persist in the rescue operations. He also urged the PLA that it should speed up the process of rescuing citizens even though soldiers would have to walk into the severely damaged areas. Wen remarked: "If walking into the most affected areas is needed, it has to be quickened even for one second earlier, thus trying our best to rescue their lives" (*Ming Pao*, May 13, 2008: A4). At the leadership level, Wen's response was swift.

From the earthquake on May 12 to the morning of May 13 at 8 AM, there were 1,100 tremors in which two reached a magnitude of at least 6 and sixteen were measured at a magnitude of at least 5. The tremors caused further damages as bricks fell down from shaky buildings and citizens became panicked in the affected regions and cities. The magnitude of the Sichuan earthquake in 2008 was comparable to the Tangshan earthquake in Hebei in 1976. Historically, the PRC had been affected by six earthquakes. Table 1.1 shows the six earthquakes in 1966, 1974, 1975, May 1976, July 1976 and August 1976. The strongest one was in Tangshan, Hebei, bringing about 242,000 deaths and 164,000 injuries.

The 2008 Sichuan earthquake affected 20 other cities and regions. Beijing recorded a magnitude of 3.9 earthquake directly shaking its city center, forcing many residents to leave their buildings for safety in outdoor areas immediately. With 88 days left for the opening of the Beijing Olympics, the Olympics Stadium in the capital city was fortunately not affected. Prior to the construction of the Olympics Stadium, construction experts had already calculated its anti-earthquake resistance ability, thus designing it to tackle any possible earthquake of magnitude 8 (*Ming Pao*, May 13, 2008: A5). Residents in Shanghai, Nanjing, Haikou, and Hong Kong also felt the tremors.

Table 1.1 Earthquakes in the People's Republic of China

Period	*Place*	*Magnitude*	*Damage*
March 1966	Hebei	7.2	Affecting 142 counties and cities, leading to 8,182 deaths and 51,395 injuries; 86 bridges collapsed and 4,000,000 houses were damaged
May 1974	Yunnan	7.1	Leading to 1,423 deaths and 1,600 injuries; and 66,000 houses were damaged
February 1975	Liaoning	7.3	Leading to 2,041 deaths and 27,538 injuries; total loss was Renminbi 8 billion
May 1976	Yunnan	7.4	Leading to 98 deaths and 2,442 injuries, with 420,000 houses being damaged
July 1976	Hebei Tangshan	7.8	Leading to 242,000 deaths, 164,000 injuries, and a loss of Renminbi 54 billion
August 1976	Sichuan	7.2	Leading to 800 deaths and injuries

Source: *Ming Pao*, May 13, 2008, p. A4.

As a matter of fact, several days before the Sichuan earthquake, some villages near the Mianzhu city in Sichuan witnessed hundreds of thousands of frogs and toads running out from their hidden places and walking on the streets, prompting public fear that this was a sign of a forthcoming natural disaster (*Ming Pao*, May 13, 2008: A7). The wandering frogs and toads occupied an area of 20 hectares, leading some residents to call up the local seismological department to ask whether there would perhaps be an earthquake. Interestingly, the forestry officials went to the villages to look into the situation and claimed that the changing weather might have brought about the normal migration of frogs and toads from one place to another. On May 7, a citizen wrote on a mainland blog and claimed that he or she was a geologist, predicting that there would be a large-scale earthquake around the central part of Sichuan and Hubei provinces. Some netizens criticized such prediction for creating panic, but after the earthquake bloggers regarded the prediction as unbelievably "accurate." A few netizens even put the blame on the officials for ignoring such warning. However, the prediction and the comments from netizens were quickly deleted after the earthquake. The "accurate" prediction, according to other netizens, was falsified. Some citizens put the blame on the inability of the Sichuan seismological officials to anticipate any earthquake; nevertheless, the officials responded that they had not detected any possible earthquake before May 12 (*Sing Tao Daily*, May 13, 2008: B5).

The power of the earthquake was calculated to be tremendous, similar to that of 256 nuclear warheads (*Ming Pao*, May 13, 2008: A7). Scientists in Taiwan assessed that the Sichuan earthquake in 2008 was about 0.5 times stronger in terms of magnitude than the earthquake that hit the Republic of China (ROC) on Taiwan on September 21, 1999, but the former's release of energy was twentyfold higher than the latter. Moreover, while the Tangshan earthquake in 1976 had been detected to occur 22 kilometers deep under the ground, the 2008 Sichuan earthquake took place at just 10 kilometers from the land surface and therefore had a much stronger and lethal impact on buildings, roads, and bridges as well as human lives. Fortunately, the 2008 Sichuan earthquake did not affect the Three Gorges Dam, which had reportedly established the capability of resisting an earthquake of magnitude 10 (*Sing Tao Daily*, May 13, 2008: B3).

On May 13, it was reported that Premier Wen was upset and angry at the speed of rescue operation over his phone call to officials. He yelled over the phone that he had to order the military to speed up the entire process of rescuing 100,000 residents being trapped in Wenchuan (*Ming Pao*, May 13, 2008: A17). He also requested that the PLA should even walk into the most devastated areas to rescue the lives of citizens and villagers. The immediate responses from the central government on May 12 are summarized in table 1.2. On May 12, when Wen was on the plane rushing to the disaster areas, he remarked that the party-secretaries and governments at all levels had to put the policy priority of saving lives on the top of the agenda, that all the damaged infrastructure would have to be recovered and reorganized, especially water supply, electricity, roads, telecommunications, and that the major construction work such as dams and railroads would have to be protected for their safety and normal operation. He added: "Comrades, we have to be calm, confident, bold and forceful in leadership in response to the disaster. I believe that under the strong leadership of the Party center and the State Council, the military and citizens must be united and we will prevail over the special and important earthquake disaster" (*Today Daily News*, May 13, 2008: B1). At the same time, the Sichuan party-secretary Liu Qibao ordered immediately after the earthquake that the party-secretaries, cadres, and government officials at all levels must mobilize party members and the masses to participate in the anti-earthquake rescue operation. In response to the appeal from Wen, the Chongqing military region sent the PLA soldiers and the People's Armed Police (PAP), together with the medical health staff members, to the affected regions.

In comparison with the government response to the 1976 Tangshan earthquake, the central government in 2008 adopted an unprecedentedly swift and open approach in handling the outbreak of the natural disaster. In 1976, the central government in Beijing did not have the capability to grasp the situation immediately after the Tangshan earthquake until one driver drove all the

Table 1.2 Immediate Response from the Central Government to the Earthquake

May 12	
14:28	The Sichuan earthquake affected provinces and cities in Hubei, Ningxia, Qinghai, Gansu, Henan, Shaanxi, Shanxi, Shandong, Yunnan, Hunan, Shanghai, and Chongqing, which all felt the tremors.
14:46	Xinhua reported that the Sichuan earthquake was of 7.8 magnitude and Beijing 3.9 magnitude.
15:12	Xinhua website publicized the topic of measures taken to deal with earthquake and citizens expressed their views through the blog.
15:50	Xinhua reported that the PLA Command Headquarters triggered the emergency crisis management mechanism and Chengdu's military region sent military personnel to understand the situation.
15:55	The party-secretary of the Chinese Communist Party (CCP) Hu Jintao issued an important directive to request immediate rescue operations to ensure the safety of citizens in the affected areas. At the same time, Premier Wen Jiabao hurried to the disaster areas to lead the rescue operations. Officials from the Ministry of Finance accompanied him and the central government deployed 2 billion yuan to deal with the disaster areas in Sichuan.
20:01	Wen Jiabao arrived Sichuan's Dujiangyan city.
Night	Hu Jintao held an emergency meeting of the Standing Committee of the CCP Politburo and listened to reports of the rescue operation. At the same time, Li Keqiang as the deputy commander of the rescue operation held emergency meetings with officials on the arrangements of the rescue work.
	The Ministry of National Land and Resources held a meeting to trigger the establishment of geological disaster response plan, setting up command leading small groups in Beijing and Sichuan and conducting investigation in the affected regions. The reason was that the post-earthquake tremors could bring about landslide, mudslide, and collapse of valleys and mountains.
	On the other hand, the transport ministry held a meeting to initiate an emergency plan of dealing with land routes, waterways, and transport systems, setting up a group of urgently repairing roads and railways that were damaged or blocked by the earthquake, trying to repair all the transport networks, and ensure smooth transport in the affected regions.

Sources: *Today Daily News* (Toronto, Canada), May 13, 2008, p. B1. *World Journal*, May 14, 2008: B1.

way from Tangshan to Beijing to report and reveal the occurrence of the disaster (*Today Daily News*, May 13, 2008: B1). However, in May 2008, it took only 18 minutes after the earthquake's occurrence for the central government to report on the incident through Xinhua news agency. Moreover, Beijing responded within one hour to issue an urgent directive from

CCP party-secretary Hu and to have Premier Wen board on his personal plane to fly to the disaster areas at once. Together with Wen were the State Council's secretary Ma Hai and Minister of Public Security Meng Jianchu. The PLA also triggered its emergency plan while the military personnel in Chengdu were immediately dispatched to the disaster zone. At the same time, citizens could grasp the most updated information through the Internet, television broadcasts, and mobile phones. Groups and enterprises also volunteered in securing donations from citizens, while other residents decided to donate their blood voluntarily. The advent of technology quickened the pace of governmental response to the earthquake in 2008.

On May 13, it was reported that 9,219 people died in eight provinces and cities, including Sichuan, Gansu, Shaanxi, Yunnan, Chongqing, Shanxi, Guizhou, and Hubei. The external communication line of Wenchuan county was simply cut off and temporarily had 57 deaths, 200 injuries and 60,000 residents missing. In Beichuan county, 7,000 people were reportedly dead, followed by 1,500 deaths in Mianzhu, and 1,000 deaths in Deyang. Beichuan had 10,000 people missing and Mianzhu's missing residents amounted to 5,500. On the first day of the earthquake, 10,832 citizens were reportedly dead and 24,460 people were missing (*Sing Tao Daily*, May 13, 2008: B1).

The PLA Command Headquarters ordered that, if the weather conditions were unfavorable to the touching down of airplanes loaded with PLA soldiers in the rescue operation, then parachuting the PLA into the affected disaster areas would be necessary (*Sing Tao Daily*, May 13, 2008: A1). The Jilin military region sent two airborne brigades to the Sichuan disaster areas, utilizing helicopters in the rescue operation. In total, 24,000 PLA soldiers and PAP officers went to Sichuan's affected areas and an addition of 10,000 soldiers were dispatched to the regions through the railway routes, apart from 2,000 police and fire service officers who were urgently sent to the disaster zones.

Many schools collapsed during the earthquake. In Wenchuan county, at least ten schools were damaged by the earthquake and some 900 students were buried, but on the night of May 12 less than one hundred were rescued. In Zhuyuan township, the collapsed Zhuchuan school had 1,700 students and 100 teachers, but only 100 students were rescued by 6 PM on the first night (*Sing Tao Daily*, May 13, 2008: B1). On the night of May 12, Beijing residents heard a rumor that the capital city would have a tremor, leading to public panic and the government's seismological department announcement that there would not be an earthquake there (*Sing Tao Daily*, May 13, 2008: B3). The Olympic torch was circulated to the province of Fujian on May 12, bringing about questions raised by some citizens that when the torch arrived at Jiangxi province, whether the process would be disrupted by any sudden earthquake. The Jiangxi provincial government quelled the rumors immediately by stressing that there would not be any transferal of the Olympic torch in Jiangxi because the tremors after the earthquake would become weaker as time passed (*Sing Tao Daily*, May 13, 2008: B3). In Zhejiang province, rumors were rife on the Internet that an earthquake of magnitude 5.7 took place, but the Zhejiang's seismological officials clarified that the Sichuan earthquake merely caused minor tremors in the province, and that the people's livelihood would not be affected. In Shenzhen and Shanghai, some 66 listed stock companies terminated their operation due to public panic (*Sing Tao Daily*, May 13m 2008: B3). The Shanghai stock exchange authorities stopped the market operation of 45 companies based in Chongqing and Sichuan. Clearly, public panic emerged and rumors were circulated in some cities after the Sichuan earthquake, prompting clarifications on the part of the provincial or central government.

Premier Wen was eager to tackle the crisis efficiently. Once he arrived at Sichuan, Wen held the anti-earthquake meeting with officials in a makeshift tent, where he made four requests: (1) grasping the time to rescue citizens as the first priority through the mobilization of more PLA soldiers and the PAP officers urgently; (2) opening the path and roads leading to the affected areas at all costs; (3) the effective coordination of health units and agencies in deploying medical staff and services, such as ambulances and pharmaceutical supplies, to the disaster zones; and (4) the full-scale resettlement of the victims by mobilizing the transport of tents so that they would be protected from rain and tremors (*Sing Tao Daily*, May 13, 2008: B3). On May 13, when Wen received the news that the rescue operation encountered difficulties and delay, he shouted to an official in the phone as follows: "I don't care about your situation now, but the order is to bring 100,000 citizens to safety" (*World Journal*, May 14, 2008: A2). He then talked to his lieutenants and military leaders in this way: "I have only one sentence: the citizens are feeding all of you and all of you must do a good job" (*World Journal*, May 14, 2008: A2).

Taiwan reacted to the earthquake in Sichuan quickly. The ruling Kuomintang or Nationalist Party sent a letter to the CCP, expressing its condolence and concern about the loss of lives in Sichuan and saying that, if necessary, it would send a rescue team to the affected regions. In reality, 4,000 Taiwan people were trapped in Sichuan province, most of them being tourists, including eight Taiwan businessmen who were missing on the first day of the earthquake (*Sing Tao Daily*, May 13, 2008: B7). The fire service department of Taiwan's Interior Ministry made a phone call to the rescue unit in the mainland, probing whether the mainland counterpart would require life detectors and sniffing dogs to rescue the victims in Sichuan. In response to the Sichuan earthquake, several mainland tours from Taiwan's Kaohsiung city to Sichuan originally scheduled from May 13 to 18 were immediately cancelled. Twenty-six Taiwan air passengers on board of an airplane that was scheduled to take off from Chengdu airport to Kaohsiung recalled that the plane had to abandon the takeoff process, and that the moving plane was shaking severely on the runway during the earthquake. Other Taiwanese working and residing in Chongqing and Shanghai reported how they witnessed public panic over the earthquake, especially in Shanghai where mainland residents were not trained to deal with earthquake, unlike the Taiwanese who had underwent such exercises back in the ROC (*Sing Tao Daily*, May 13, 2008: B7).

The number of deaths on the night of May 13 reached 12,012 while 9,404 people were buried and 7,841 others were reportedly missing (*Sing Tao Daily*, May 14, 2008: A1). The unofficial number of deaths amounted to 26,000. In Yingshou county, the communication of 60,000 citizens with the outside was cut off while the reservoir nearby would have the danger of collapse. Another devastated area was Beichuan county with 7,000 deaths and 20,000 people missing, while a collapsed school trapped at least 1,000 students and teachers. Twenty-four hours after the earthquake, the Wenchuan party-secretary Wang Bin could contact the authorities outside his county but he broke down into tears at once over the phone discussions. On May 14, it was reported that 10,000 people died in Yingshou county and only 3,000 survivors stayed there (*Sing Tao Daily*, May 14, 2008: A1). Another report said that 18,800 residents had died in Yingshou and only 2,300 people survived (*Ming Pao*, May 14, 2008: A5). About 70 percent of the bridges and roads leading to Yingshou were destroyed, thus generating difficulties in the rescue operation.

The military and police walked for ten hours into the affected regions amidst rainfall. The Sichuan military region's commander Xia Guofu led the elite troops to arrive at Wenchuan county at noon on May 13, while the same region's deputy commander led 100 commandos and 120 emergency staff officers to reach the destination simultaneously. Another 200 PAP police led by chief military advisor Wang Yi walked for 90 kilometers and eventually arrived at the Wenchuan county's capital city. On the way to their destination, all the bridges collapsed and roads severely damaged and the military officers, police, and medical staff had to distribute urgent supplies to the earthquake victims. A military brigade from Jilin had 100 officers being airlifted by a large passenger plane of the China Airline and parachuted into the earthquake-wrecked areas (*Sing Tao Daily*, May 14, 2008: A1). Moreover, over 60,000 PLA paratroopers and special armor units were parachuted and sent to the affected regions, which were severely destroyed and which hindered the use of any large-scale rescue instruments. As such, the military soldiers and police used their hands to remove the debris and bricks to rescue the earthquake victims. During the post-earthquake tremors, they stopped the rescue operations. It was calculated that after the earthquake, there were 1,950 tremors in which three reached a magnitude of 6 and 14 measured at the magnitude of 6. Thirty-eight military aircrafts were utilized and flew 120 times in two days' time, while 4,000 airborne troops went to rescue the earthquake victims. The swift response of the PLA could be seen. On May 14, 5,000 fire services officers were dispatched to Sichuan from Beijing, Tianjin, Shanghai, and Guangdong to bring various rescue equipments, including torches, stone-breaking tools, cranes, and telecommunications facilities.

GOVERNMENTAL RESPONSE ON MAY 14

Premier Wen expressed his deep emotions on May 14 when he went to inspect the situation in Beichuan county. He told the earthquake victims and citizens that 100,000 PLA soldiers, PAP paramilitary police, and civilian police were now rushing to different affected districts. Wen added among the 100,000 mobilized rescue troops and police, 60,000 came from the PLA, 25,000 from the paramilitary police, and 10,000 from civilian police and fire services officers. Some 9.1 tons of telecommunications systems, medical supplies, and quilts were transported to the disaster zones (*Ming Pao*, May 14, 2008: A1). According to Wen, the Party and the government tried their very best to implement the rescue operation. He expressed his gratitude to the support and concern from American citizens and the Hong Kong comrades, including American volunteers and Hong Kong reporters who rushed to the scene to help the rescue and report the disaster respectively. Wen held the hands of a girl and added: "Do not cry! Put your heart at ease because the government is managing you, other victims, your lives and your study. You will be the same as with the situation at your home" (*Ming Pao*, May 14, 2008: A1). He hoped that the girl and the victims would persist, but the girl broke down into tears and nodded to the premier. Wen was moved by her reaction and told the officials beside him to take special care of the orphans in the earthquake. He shouted tearfully to the children: "I am uncle Wen Jiabao and children you must withstand the situation" (*Ming Pao*, May 14, 2008: A1).

On May 14, Premier Wen went to a collapsed school in Dujiangyan where rescue workers used their hands to try digging out 300 children and teachers being trapped under the debris and rubble. When he noted that babies lacked milk powder and water supply was deficient, Wen made five requests: (1) the military had to go into the city from north and south immediately; (2) all the roads had to be repaired; (3) the circumstances of disaster areas had to be found out; (4) every government department had to think about how to get the logistical supplies to be sent to the affected areas; and (5) the seismological departments and officials had to study the trends of the earthquake swiftly and scientifically (*World Journal*, May 14, 2008: A2). Dujiangyan was a historical and ancient city with 49 nationally protected cultural heritage sites. Although the cultural heritage sites were not damaged, the damage done to the buildings was severe. Moreover, the heavy rainfall and post-earthquake tremors trapped the rescue operation teams inside the city and helicopters could not touch down onto their destinations (*World Journal*, May 14, 2008: A2).

The city of Dujiangyan, originally having a population of 600,000 residents, was only 50 kilometers away from the epicenter and became one of the most devastated areas. At least three schools collapsed in the city, including two secondary schools and one primary school. Eighty percent of the city's buildings were destroyed and eighty percent of its residents also became homeless. Some parents of the trapped and dead children in the collapsed schools shouted loudly at why the local government had made "tofu" buildings, which referred to schools constructed with substandard concrete because of bureaucratic corruption. One parent said: "This collapse was obviously a result of substandard construction, thus victimizing our children" (*Ming Pao*, May 14, 2008: A4). About 400 students and teachers were trapped in a collapsed school, but only 20 of them were rescued. Other dead bodies dug out by rescue workers looked terrifying in their appearance, leading to painful cries among parents almost everywhere. Some parents shouted tearfully that they had to find and see their dead children. Seven to eight truckloads of PLA soldiers went to Oi village, but due to the fact that large-scale rescue cranes were insufficient, only three tractors removing soil and rubble could be mobilized. Some parents were so upset that they put their anger on the rescue workers, blaming that they were too late to arrive at the scene. Nevertheless, the rescue workers took 12 hours to walk through a distance of 10 kilometers—a difficult process of rescue operation. Another parent said angrily that the entire city's children were dead, while other parents hugged their dead children and shed their tears painfully" (*Ming Pao*, May 14, 2008: A4). Parents of the dead children in other places also put their blame on the "tofu" schools, such as Chongqing's Liangping county where they pointed the fingers at the substandard construction of schools (*Ming Pao*, May 14, 2008: A7).

Two other devastated zones were Wenchuan county and Beichuan county, the former originally having a population of 161,000 and the latter with 106,000 residents. Wenchuan was located precisely at the epicenter whereas Beichuan was 80 kilometers away from the center of the earthquake. Beichuan witnessed many victims, some of whom had their legs cut off by rescue workers in order to save lives effectively. In one school, almost 1,000 people were buried and some trapped victims groaned amidst debris and rubble. The post-earthquake tremors exacerbated the deaths and casualties as rocks rolled down from mountains. On May 14, it was reported that Beichuan had over 7,000 deaths and 20,000 people missing. The county no longer had electricity and water supplies, prompting residents to walk to Mianyang and Chengdu cities. Mianyang was about 160 kilometers away from Beichuan and 100 kilometers from Chengdu. In Wenchuan county, one-third of its buildings collapsed and 200 PLA soldiers arrived at the scene on May 14 to rescue the victims. About 200 more PLA officers were dispatched to Wenchuan through the water route while the necessary supplies were airlifted to the region. Premier Wen went to both Beichuan and Wenchuan counties on May 13 to inspect the circumstances. Located at the southeast of Wenchuan, the Mianzhu city was another devastated area after the earthquake. On May 14, 2,000 people died and 10,000 citizens were injured (*Sing Tao Daily*, May 14, 2008: B4). After the earthquake, due to shortage of supplies, residents in Mianzhu needed to use 600 yuan to buy nylon cloth to construct their own tents so as to protect themselves from the rainfall and they also needed 50 yuan to buy a bucket of water. In Mianzhu's Tianchi town, 6,000 people were working in a phosphorous coal mining operation when the earthquake took place and there was no report saying that any miners were eventually rescued (*Toronto Star,* May 16, 2008: A21). Circumstances were unfavorable to the survivors.

In Chengdu city, 400,000 residents went outdoors to sleep over the first night of the earthquake due to their fear of the post-earthquake tremors, but the numbers declined as public panic gradually subsided. Chengdu had a population of 10 million. Its airport was temporarily closed down but reopened on May 13, including all the railway lines connecting the city to other places. The 86 pandas in Sichuan province were not hurt, so as the 68 pandas residing in Shaanxi province.

Although many citizens were panicking during and after the earthquake, some citizens formed volunteer groups to rescue the victims. In the cities of Chengdu and Chongqing, 500 residents formed self-help groups through organization in their work units and Internet discussions (*Sing Tao Daily*, May 14, 2008: B1). One netizen wrote: "My profession can contribute to the post-earthquake reconstruction work because I am an architectural designer. I can help victim families to design their houses. If there are any volunteers, please inform me and let us help the victims together" (*Sing Tao* Daily, May 14, 2008: B1). After Premier Wen made tearful appeal to the military and police to rush to the affected zones before the midnight of May 13, some citizens expressed their emotions through Internet blogs and discussions, saying that the Chinese nation must be able to prevail over all difficulties. In Guangxi and Wuhan, many citizens immediately went to donate their blood to help the victims in the earthquake. While the government of the Hong Kong Special Administrative Region (HKSAR) and the local business sector secured HK$3.5 billion on May 13 to help the earthquake victims in the mainland, the PLA stationed in the HKSAR also raised HK$1.5 million donation (*Sing Tao Daily*, May 14, 2008: B2). Similarly, the government of the Macao Special Administrative Region (MSAR) immediately raised

MOP$1.1 billion for the victims of the Sichuan earthquake. The immediate responses from the ordinary citizens and local governments, including the HKSAR and MSAR, to support and help the victims demonstrated the solidarity of the Chinese in coping with the 2008 earthquake.

In response to Premier Wen's appeal for swift action, the PLA mobilization was prominent on May 13. The Ministry of Defense announced that 20,000 soldiers arrived at the affected zones while 30,000 others were rushing to the scenes through railways, roads, and aviation routes, including 6,000 elite paratroopers (*World Journal*, May 14, 2008: A6). In the Chengdu military region, 30 helicopters and 8,600 PLA soldiers as well as 5,000 PAP police conducted the rescue operation at once. At the same time, the airborne division sent 22 planes and the aviation department mobilized 9 transport planes to dispatch 11,000 soldiers directly to the disaster-affected areas. About 15,600 reserves soldiers and officers from Sichuan and Gansu military regions as well as the Chongqing police region were also mobilized to open the paths and roads leading to the disaster areas, including 67 medical teams and 1,500 medical staff members rushing to Sichuan. On May 13, the Chengdu airport was filled with tents, blankets, food, and medical supplies from Jilin, Shenyang, Guangzhou, and Lanzhou military regions (*World Journal*, May 14, 2008: A2). On the morning of May 14, the PLA already airlifted 11,420 soldiers into Chengdu—an unprecedented operation in terms of the number of aircrafts and the frequency of flights on a single day (*World Journal*, May 14, 2008: A6).

On May 14, the Standing Committee of the CCP Politburo held a meeting presided by party-secretary Hu Jintao over the progress and details of the rescue operation. The meeting listened to the reports from the Chief Command Headquarters under the State Council on the progress of the rescue work. The Politburo meeting made a judgment saying that "all kinds of relief and rescue operations are proceeding systematically and the social order remains largely stable" because of "the strong support of every government agency, the unselfish societal assistance, the emergency help of local party-secretaries and governments, and the frontline work of the PLA and PAP under the leadership of the Party Center, the State Council and the Central Military Commission" (*Today Daily News*, May 15, 2008: B5). The meeting also decided that all local governments and agencies concerned would have to work diligently and in solidarity so as to "win the battle over the anti-earthquake disaster relief work" (*Today Daily News*, May 15, 2008: B5). The leadership of the CCP was emphasized in the Politburo response to the progress of the relief operation.

The CCTV responses were also swift, sending reporters to cover the earthquake regions in an unprecedentedly open and transparent manner. The broadcast of CCTV coverage of the 2008 earthquake was made available to the overseas Chinese in Canada through the Fairchild Television, especially on the first several days of the disaster (Personal observation, May 12 to May 15, Toronto, Canada). The CCTV sent reporters to look into various affected regions, allowing the audience to have access to the tragedies. In one of its reports, the CCTV revealed that some helicopters were too large as to descend and touch down into difficult terrains—an unprecedented openness on how the PLA encountered its difficulties in the rescue operation. The CCTV coverage also showed poor visibility and airlifting difficulties of helicopters that tried to download logistical supplies. Surprised by the audience inside and outside China, the CCTV also broadcasted the remarks of earthquake survivors who were critical of the slowness in the rescue efforts of local government officials—an unprecedented show of openness of the official CCP mouthpiece (*World Journal Weekly*, May 25, 2008: 18). In a sense, the openness of the official CCTV demonstrated the confidence and determination of the PRC government to cope with the crisis in a collective, united, and rapid manner.

The mainland responses were prominently rapid. Guangdong province collected donations of relief logistics amounting to 1 billion yuan, while other provinces such as Heilongjiang, Guizhou, Inner Mongolia, Ningxia, Hubei, and Jilin over 60 billion of monies and relief supplies within the first three days after the earthquake (*Today Daily News*, May 15, 2008: B5). Banks, automobile companies, and public and private sector organizations donated milk products, logistical supplies, and monies immediately. All the supportive measures were poured into the Sichuan province—a testimony to the unprecedented societal and provincial reactions after the earthquake.

The people in the Republic of China (ROC) on Taiwan reacted to the Sichuan earthquake swiftly and warmly. Taiwan Plastic donated 60 million yuan while many other Taiwan business people donated money, food, and medical supplies (*World Journal*, May 14, 2008: A4). The ROC Red Cross donated US$300,000 to support the earthquake victims and logistical supplies were immediately sent to Chengdu. The ROC president Ma Ying-jeou appealed to the Taiwan people to support the earthquake victims; he donated NT$200,000 to express his sympathy and support for them (*World Journal*, May 14, 2008: A4). Other Taiwanese religious organizations, including the Buddhist, collected blankets from citizens and donations from their followers. One Taiwan charity foundation in Shanghai and Beijing organized 17 volunteers to go to Sichuan at once, while 60 volunteers in Taiwan also organized themselves to participate in the rescue operation in mainland China. Although Taiwan remained politically separated from the PRC, some Taiwanese took immediate action to demonstrate their deep concern and support of the earthquake victims.

Seismologist Geng Qingguo, who had accurately predicted the 1976 Tangshan earthquake's occurrence, said that there was an observed pattern of the relationship between the 2008 Sichuan earthquake and droughts. The epicenter in Sichuan experienced droughts from 2005 to 2007, a phenomenon demonstrating a historical pattern that droughts lasting from one to three and half years continuously would signaled a forthcoming earthquake (*Ming Pao*, May 14, 2008: A23). In July 1984, Geng argued that China's northwest regions experienced 69 earthquakes from 231 A.D, when Emperor Qin ruled ancient China, to 2202 A.D., and that the occurrences amounted to 97 percent of the earthquakes in Chinese history (*Ming Pao*, May 14, 2008: A23). There were 16 earthquakes that took place one year after droughts, 15 occurring two years after droughts, 16 breaking out three years in the wake of droughts, and 9 erupting three and half years following droughts. Geng had predicted the Tangshan earthquake in 1976, but the seismological authorities were concerned about public panic in the capital city of Beijing and therefore did not issue any pre-earthquake warning (*Ming Pao*, May 14, 2008: A23). Geng appealed to the government for learning a lesson from the historical pattern of the dynamics between droughts and earthquakes. He added that, apart from droughts, other short-term indicators of earthquakes included whether the air pressure was persistently low in consecutive days, whether temperatures stayed high or low continuously, whether a sudden heavy rainfall occurred in several decades, whether toads and frogs suddenly migrated, whether other animals had abnormal behavior, whether the water level under the land surface moved abnormally, and whether the light reflections on the land surface showed drastic changes. Although Geng's analysis appeared to be insightful, other seismologists tended to be cautious and remarked that it was difficult to predict the occurrence of earthquakes (*Ming Pao*, May 14, 2008: A23).

In comparison with the 1976 Tangshan earthquake, Chinese seismologists remarked that the 2008 Sichuan earthquake rendered rescue work very difficult because of the mountainous surroundings of the affected regions, especially Wenchuan. Although the magnitude of the Tangshan earthquake was the same as the Sichuan one, the latter had more post-earthquake tremors that made it difficult for rescue operation to proceed smoothly. Wenchuan, which had an earthquake in 1657, suffered severely from such tremors (*Sing Tao Daily*, May 14, 2008: B4). During the Sichuan earthquake, Wenchuan's villages located on the mountainous areas were annihilated immediately while its reservoirs posed tremendous dangers to the lives of survivors after the earthquake as they faced the danger of collapse.

Some Chinese scientists asserted that while the 1976 Tangshan earthquake led to the government's mobilization of students to constantly examine the water level of wells for several years, such predictive work related to earthquake was later neglected (*Ming Pao*, May 18, 2008: A6). Due to the difficult terrain and valley-shape of Wenchuan, where many roads were constructed along rivers, any earthquake would make it difficult for rescue operation, not to mention large rescue machineries that could get into the disaster regions immediately after the 2008 Sichuan earthquake. Compounding the complex landscape in Wenchuan was the rapid rural-urban development in the PRC from the mid-1970s to 2000s, when houses and buildings in Wenchuan were constructed without serious anti-earthquake considerations. The Sichuan earthquake therefore would prompt the government authorities to reassess carefully whether some earthquake-prone regions should be re-populated and reconstructed.

GOVERNMENTAL RESPONSE ON MAY 15

On May 15, while Premier Wen went to Qingchuan county to inspect the situation as 800 people died there and 300 were buried, the dam surrounding a reservoir at the Taiping hydroelectricity station, which was 9 kilometers away from the Yingshou county, had the danger of collapse, thus endangering the lives of many survivors. The Sichuan water and electricity company assessed that 110 hydroelectricity station staff members and 30 residents were trapped in the station around the water dam, which would likely collapse because of a rapid increase in the water level after the earthquake. As such, the survivors would have to be rescued through the air route. The Taiping hydroelectricity station belonged to one of the nine hydroelectric stations that were constructed along the Minjiang in 2003, when some geological experts had already warned that building too many stations would encounter grave dangers in case of any earthquake, especially as the several reservoirs there would cause great damage in crisis situation (*Today Daily News*, May 16, 2008: B2). Their views were not heeded, however. As Zipingpu reservoir was in the danger of collapse, 2,000 PLA soldiers were sent there to prepare rescuing the workers and residents, while excessive water in the reservoir was urgently released so as to minimize the pressure on the dam (*Today Daily News*, May 16, 2008: B2). Finger-length cracks scarred the top of the dam, leading to the possibilities of landslides. The Ministry of Water Resources issued a notice to check reservoirs nationwide, whereas the economic agency reported that at least 400 dams were damaged (*Toronto Star*, May 15, 2008: A3).

Maintaining the health conditions in affected regions was of utmost importance. The World Health Organization warned that the Sichuan earthquake might have the side effect of producing communicable diseases, while Russian health experts told the PRC officials that Russia would be delighted to provide vaccines preventing against any rat diseases (*Today Daily News*, May 19, 2008: B5). In fact, the Chinese health officials reacted to the Sichuan earthquake promptly. On May 15, about 10,000 health officers arrived at Wenchuan. Among the 10,000 health workers, 1,200 volunteers came from different provinces and cities; 1,755 were sent from heath agencies in various provinces; 5,000 from Sichuan province; and the rest were coming from the PLA and PAP forces (*Sing Tao Daily*, May 16, 2008: B5). The Guangdong health team also sent two psychological counseling groups to deal with the psychological shocks suffered by the earthquake victims. The Ministry of Health announced that six measures were implemented to stop the outbreak of any infectious diseases, including the protection of the safety of drinking water and food, the adoption of anti-viral work, the injection of preventive medicine into the bodies of survivors and residents in affected regions, the proper arrangements of corpses, the close monitoring of infectious diseases, and the importance of hygiene and health safety of the rescue workers (*Sing Tao Daily*, May 16, 2008: B5). The Ministry of Agriculture sent work teams to the affected regions in Sichuan and other provinces where 1.2 million animals and livestock died, necessitating swift action on the part of agricultural officials to deal with the dead animals for the sake of preventing the outbreak of other diseases (*Today Daily News*, May 18, 2008: B7).

By May 15, 9 brigades of the PLA went into Sichuan, especially Wenchuan county and its nearby six cities, where altogether 20 million residents were affected by the earthquake. Several thousand PLA brigades walked into all the affected regions, shouldering the logistical supplies and rescue tools. They also arrived in 58 isolated counties, successfully reaching the regions that were cut off from any immediate assistance. On the night of May 13, Wenchuan county's highway number 317 opening to the outside regions was eventually reopened. The PLA air force sent 100 helicopters into Wenchuan county, apart from 80 other planes. Overall, 150,000 PLA soldiers, PAP police, and civilian police were already mobilized to different affected regions in Sichuan province, while 64,040 injured citizens were taken care of by various hospitals in the province (*Today Daily News*, May 16, 2008: B3). The Ministry of Health in Beijing announced that infectious disease had not broken out in the affected regions and that it had swiftly dispatched 5,000 health personnel to the scenes. All kinds of medicine were sent to the affected regions, including antibiotics, breathing instruments, and pain suppressants. Nevertheless, some children in the affected regions developed skin disease due to air and water pollution. Health workers in the affected regions sprayed anti-decay medicine onto the dead bodies to prevent the eruption of infectious diseases, although many other dead corpses were left on the roads and streets without sufficient care. Overall, the governmental responses to the post-earthquake rescue operation were prompt and stemmed the outbreak of infectious diseases.

Although the PLA was fully mobilized to rescue the victims, some citizens were so eager to leave the disaster areas that they competed with each other to get onto the rubber rafts prepared by the PLA in Baihua county, which was severely damaged by the earthquake. In Dujiangyan city where many residents died, a funeral service company not affected by the earthquake witnessed the families of victims buying and grasping candles and paper money for the sake of honoring and remembering their dead relatives (*Today Daily News*, May 16, 2008: B3). Although many private and government vehicles rushed into the city to carry the dead bodies away and provide the necessary logistics respectively, the long queue of vehicles led to traffic jams that ironically slowed down the rescue operation.

On May 15, the PRC government agreed with the idea that rescue teams from Taiwan and Japan would be allowed to arrive at the disaster regions in Sichuan. The Foreign Ministry announced that Japan would send a professional rescue team to the province, and that it welcomed Taiwan's Red Cross to send another 20-member rescue team to Sichuan. The ROC also sent four flights to Sichuan to evacuate 1,000 Taiwanese tourists who were trapped in the province. The Japanese government formed a rescue team of 40 to 50 members, who came from the fire services department, police, and marine security agency (*Today Daily News*, May 16, 2008: B3). As Japan had rich experiences in dealing with earthquakes in the past, its rescue team could benefit the Chinese rescue operation in Sichuan.

Following Taiwan and Japan, other countries also lent their helping hands. While the United Nations Central Emergency Response Fund provided US$7 million to China for immediate relief, the European Union gave 2 million euros to China as disaster relief. Other countries like Singapore, Russia, and South Korea also sent rescue operation teams to the PRC (*World Journal*, May 17, 2008: A6). The spokesman of the State Department of the United States government, Sean McCormack, remarked that the US Pacific fleet would use C-17 transport planes to airlift tents, blankets, food, and logistical supplies to the PRC, and that the US satellites found that there was no leakage in any mainland Chinese nuclear weapons study facilities (*World Journal*, May 17, 2008: A6). On the other hand, the American health secretary Mike Leavitt who visited Beijing revealed that the American side exchanged information with the PRC officials on how to prevent the spread of infectious diseases after the 2005 Hurricane Katrina. Obviously, intergovernmental exchange of information on nuclear safety and disease control contributed to the joint efforts made by both China and the United States to respond to the 2008 Sichuan earthquake.

COMPARING PREMIER WEN JIABAO'S ROLE IN WENCHUAN EARTHQUAKE WITH PREMIER ZHOU ENLAI'S ROLE IN THE 1966 XINGTAI EARTHQUAKE

It must be noted that while Premier Wen was unhappy with the performance of the PLA in the 2008 Wenchuan earthquake, his role was a far cry from the late Premier Zhou Enlai, who went to inspect the affected regions shortly after the outbreak of the Xingtai earthquake in Hebei province on March 8, 1966. The Xingtai earthquake killed 8,000 citizens and injured 40,000 others (Zhou, 1997: 78). Immediately after the earthquake, Zhou reported the situation to the CCP Chairman Mao Zedong, who asked the State Council to take the full responsibility of tackling the rescue and relief operation and also ordered the PLA to assist the rescue work (Zhao, 2008: 4). In response to Mao's directive, Premier Zhou Enlai held an emergency meeting in the State Council to formulate the rescue plan, which was then submitted to Mao for his approval (Zhao, 2008: 4). Zhou then met officials from the national science commission, the Chinese Academy of Science (CAS), seismic authority, geological department, and oil and gas agencies, designating the CAS to take the lead in inspecting and monitoring the earthquake's development closely (Zhou, 1997: 79). After he met with the State Council officials, Zhou decided to inspect the situation in Xingtai while ordering the PLA near the Xingtai area to rush to the earthquake-affected regions and asking the air force to prepare two airplanes for him and other officials to fly to Xingtai immediately. When Zhou arrived at the scene and inspected the circumstances, he expressed his gratitude to the PLA work and told victims that Mao had sent him to take care of their situation (Zhou, 1997: 78). In Xingtai, Zhou listened to the reports from officials and decided to set up a command headquarters led by the PLA because the situation was extraordinary and therefore the PLA-led rescue operation would be more efficient (Zhou, 1997: 79). The PLA was also responsible for the burial work of the victims while the other departments, including medical teams, took measures to prevent any outbreak of infectious disease. One of the most impressive actions of Zhou was that he went to inspect the earthquake-plagued regions for several hours without lunch on March 10; when he was thirsty and delivered a dirty bowl full of water, Zhou drank the water at once and left a deep impression on many residents (Wang, 2005: 119).

Several differences between Wen's visit to the Wenchuan region and Zhou's visit to Xingtai could be seen. First and foremost, Zhou Enlai was a unique premier whose authority was not simply based on his civilian position but also stemmed from his long-time participation both in the PLA as a political commissar in 1931 and in the Central Military Commission as a vice-chairman in 1933. Unlike Zhou who commanded tremendous authority as a civilian-military leader, Wen was a civilian leader without military background. Because of their different political career paths, Wen's public displeasure with the PLA "sluggishness" was contrary to Zhou's praise and support of the PLA's diligent rescue work. Moreover, the constitutional role of the PRC premier is to be the leader of the government and administrative apparatus. As such, Wen could not really command the military in the constitutional sense. Even though Zhou Enlai ordered the military to implement the rescue operation of the Xingtai earthquake, he skillfully emphasized to the citizens that he was entrusted and empowered by Chairman Mao to do so. Unlike Zhou who stressed the hierarchical relationship with Mao, Wen appeared to be portrayed in the official Chinese media as a premier who followed the tradition of Premier Zhou in visiting the earthquake regions immediately in 2008 (*Xinhua*, May 23, 2008). Unintentionally, the hierarchical relationship between Wen and President Hu Jintao was played down. Yet, it was Hu who was also the chairman of the Central Military Commission. Viewed from these perspectives, it was natural that some PLA officers were unhappy about Wen's public criticisms of their work, especially when the rugged and mountainous terrains of the Wenchuan regions made rescue operations relatively difficult.

CONCLUSION

The governmental responses to the May 12 earthquake were swift, characterized by the immediate action on the part of Premier Wen, the CCP's Politburo, the State Council, the PLA, the PAP, and other provinces as well. If responsiveness is a crucial indicator of crisis management after an earthquake, the 2008 Sichuan earthquake did mark the high degree of governmental responsiveness in the PRC at the central, provincial, and local levels. Although Premier Wen vented his anger at the PLA's relatively slow responses, the unprecedented scale of disruption incurred by the earthquake in Wenchuan and the rugged terrains made the PLA difficult to expedite the rescue operation. Compared with the late Premer Zhou Enlai who commanded tremendous respect and status among the PLA officers, Wen was more a civilian leader rather than a combined civilian-military commander as with Zhou. Still, the swift responses of Premier Wen, the State Council, and the PLA paved the way for the operational success in the rescue of the victims immediately after the Wenchuan earthquake.

Chapter Two

The Response of Interest Groups, Citizens, and Tibetan Activists

Three days after the May 12 earthquake in Sichuan, interest groups and citizens organized themselves in order to rescue the victims. This chapter examines how interest groups and citizens formed a vibrant civil society in response to the outbreak of the earthquake from May 16 to May 22—an unprecedented phenomenon in the PRC's political and social history.

THE SOCIETY'S RESPONSE

On May 16, it was reported that some survivors, including school children buried under the rubble and debris, tried their best to save themselves by various means, such as singing songs that boosted their morale while waiting for rescue workers and shouting continuously so that they would be rescued from hidden places (*Today Daily News*, May 16, 2008: B5). The medical staff members who were sent to rescue the victims were reportedly abandoning any rest but to conduct their rescue operation continuously for at least 42 hours. Volunteers flooded into different cities, such as Deyang and Shifang, to dig out the school children by their hands and any tools that they could find. They assisted the firemen and rescue workers in saving the lives of 30 students although 3 teachers had died amidst the earthquake (*World Journal*, May 16, 2008: B2). One brave teacher reportedly rushed into a classroom to rescue 52 children from a nearly collapsed kindergarten in Pengzhou city (*Sing Tao Daily*, May 19, 2008: B3). Some actors and actresses formed volunteer rescue groups and went to Wenchuan to care for the victims, providing them with the psychological and practical support (*Today Daily News*, May 19, 2008: B5). Thousands of artists performed for the CCTV's special programs in response to the earthquake, raising 1.5 billion yuan for the victims and survivors. Orphans in the Tangshan earthquake also took the initiative to donate 30 million yuan to the earthquake victims, including donation aimed at rebuilding schools for the Sichuan children. In Chengdu city, donors quickly opened a school for 166 children affected by the earthquake (*Today Daily News*, May 19, 2008: B6). Suddenly, the Chinese civil society full of volunteers sprang up in the assistance of the earthquake victims and survivors.

Other volunteers helped the PLA soldiers to bury the dead bodies. In principle, the PLA soldiers in Shifang city dug out a burial site for each dead body, but due to the fact that the temperature of Shifang rose to 28 degrees Celsius and that more dead bodies piled up, they had no choice but to use the mass burial approach (*World Journal*, May 16, 2008: B3). Simple tribute was paid by the relatives of the victims before the dead bodies were sent to crematoriums and many dead bodies were not buried by using coffins (*Ming Pao*, May 16, 2008: A6). Volunteers from the mainland and Hong Kong continued to carry logistical supplies to the survivors, but along the way they saw lots of dead bodies wrapped inside plastic bags under the hot weather, while health workers continued to use antiseptic chemicals to spray on the dead corpses to stop any outbreak of infectious diseases. In Deyang city, over 1,000 dead bodies were burnt on the fourth day after the earthquake (*Ming Pao*, May 16, 2008: A6).

Moving stories involving individuals and volunteers abounded. Citizens helped each other after the earthquake, while many bought their own air tickets to fly to Sichuan from other provinces and cities, joining the volunteer groups to assist the earthquake victims (*Ming* Pao, May 19, 2008: A7). Many cities were willing to adopt the orphans resulting from the earthquake in Sichuan. Many relatives ignored the post-earthquake tremors and went into dangerous zones to look for their loves ones. A photo showing a man carrying his dead wife on a motorcycle in Mianyang was published in the British newspaper *Daily Telegraph* and reprinted in many Chinese newspapers and websites (*World Journal*, May 18, 2008: B1). He wanted to have the dead body of his wife cremated in a funeral parlor. Some trapped victims were eventually rescued by workers even six to eight days after the earthquake—phenomena that were described as miracles by many reporters. One child named Yang Jianhua wrote in her diary in both Chinese and English and she revealed how volunteers helped her during the earthquake. She wrote:

> The earthquake destroyed our hometown. We have nothing now. Many people have no home. Many children become lonely. This happened on May 12, 2008. At that time, we were just having the first class in the afternoon. The earthquake came too quickly and we had no such expectation. It was very fearful. When I saw the dead people, my heart was hurt. I cried and later I had

> no crying sound except for my tears. After the earthquake, we had nothing. It was the Changhong company that gave us help. Its boss and workers gave us food, water and clothes. The most important thing they did for us was to rebuild the Beiyizhong School. They give us help and we will never forget (*Ming Pao*, May 19, 2008: A19).

In another very touching scene, a disabled man who lost one of his legs used his hands to climb to the Guangzhou street station that received donations from citizens, donating his 30 yuan to the relief workers (*Zhonghua Tangshou*, May 30, 2008: 18). Beggars in the provinces of Guangzhou, Anhui, and Jiangsu donated their monies from begging. Observers said that the Chinese civil society suddenly emerged in the wake of the Wenchuan earthquake, but it was a Chinese-style society supportive of the PRC government's measures to rescue the earthquake victims and begin immediate reconstruction work.

The PRC Federation of Chinese Women Associations immediately set up a children relief fund to help those orphans losing their parents in the earthquake. The federation also publicized its account number nationwide so as to receive donations from citizens in other provinces and cities. A blogger named Angela in Changsha appealed to citizens through the Internet to donate money for the victims and she said that she had been moved by Premier Wen's contacts with the orphans. Three hundred bloggers responded to her call immediately and the Department of Civil Affairs in Sichuan province reacted to the situation quickly by opening two telephone hotlines so that citizens interested in adopting orphans from the earthquake could ask questions relating to the details of adoption (*World Journal*, May 16: 2008: B4).

The response of many mainland citizens was calm and swift. Five days after the earthquake, the Sichuan market returned to normal and kept supplying rice, flour, pork, eggs, beans, water, and other necessities, while other provincial markets remained operative. The agricultural officials ensured that the price of goods and vegetables in Sichuan returned to the level before the earthquake, while 70 percent of the earthquake survivors in affected regions and 70 percent of the enterprises in Sichuan province had water supply (*Today Daily News*, May 18, 2008: B6). Although about 1.4 million people in affected regions still did not have water supply, the calm responses from the government and citizens at the provincial and local levels were noteworthy. The Chengdu education department provided temporary relief of 5,000 yuan to each parent who lost their children, while Premier Wen announced that the State Council provided rice and 10 yuan as daily subsidy to each victim of the earthquake disaster and also 5,000 yuan as relief money to each family member who lost his or her close relatives (*Today Daily News*, May 18, 2008: B6). The Fudan University in Shanghai sent psychologists to assist the ethnic minorities, such as the Tibetans and the Qiang. The Eastern

Airline responded to the earthquake by providing 46 civilian planes and 6 helicopters in the relief operation, and extracting 10 yuan from the airfare of every air ticket as part of the donation efforts helping the poor and the needy in the earthquake (*Today Daily News*, May 18, 2008: B6). In Guangdong where hundreds of thousands of migrant workers from Sichuan were working, the labor bureau issued a notice to all enterprises and urged them that they should pay the salaries of the guest workers earlier while providing the necessary logistical support.

One of the most impressive provincial responses was from Yunnan. About 4,857 volunteers from the public and civil servants were formed and sent to Sichuan immediately (*Today Daily News*, May 26, 2008: B1). On the other hand, 2 million CCP members in Yunnan province were mobilized to pay special party dues in which the monies were allocated as donations supportive of the earthquake survivors. Furthermore, Yunnan province welcomed the arrival of 1,000 injured citizens in the Wenchuan earthquake for medical treatment. Many other volunteers organized themselves to go to Sichuan, while public donations continued for the sake of acquiring more practical and financial support for Sichuan's rapid recovery and reconstruction later. The provincial government also organized the ambulance service personnel, fire services, and medical teams to be sent to Sichuan, demonstrating the warm response from Sichuan's neighboring province.

RESPONSES FROM HONG KONG AND TAIWAN

Volunteers from Hong Kong arrived at Sichuan quickly and joined the rescue operation. The Hong Kong branch of Doctors Without Borders sent doctors to Sichuan immediately, whereas the Hong Kong Salvation Army organized social workers and volunteers to go to the affected regions (*Ming Pao*, May 16, 2008: A20). The late Run Run Shaw, who passed away in 2013, donated 100 million Hong Kong dollars to the relief and rescue operation, while other agencies poured their monies and donations into the PRC, including the Hong Kong Chinese Enterprises (HK$60 million), the Hong Kong Jockey Club (HK$30 million), the Chiang family (HK$10 million), Stanley Ho Hung-sun (HK$10 million), Ronald Arculli (HK$5 million), Tung Chee-hwa (HK$5 million), the Hong Kong Democratic Party (HK$1 million), and the Social Democratic Front (HK$95,000). The Hong Kong Fire Services Department sent a team of 20 rescue workers who arrived in Mianzhu to conduct the rescue operation immediately. Members of the law-making Legislative Council of Hong Kong wrote a joint letter to Premier Wen, expressing their condolence of those Chinese citizens who died in the earthquake. The University of Hong Kong donated HK$160,000; the Hong Kong Baptist University HK$130,000, the Hong Kong University of Science and Technology HK$100,000, and the Hong Kong Institute of Education HK$100,000 (*Ming Pao*, May 17, 2008: A20). On May 19, the Liaison Office—the PRC government's representative agency in Hong Kong—reported that it received HK$400 million donations, including Li Ka Shing's Foundation's HK$30 million and numerous milk powder and medicine that were then transferred to Sichuan province (*Today Daily News*, May 19, 2008: C3). The quick response from Hong Kong was so prominent that the PRC president Hu Jintao expressed his gratitude to the warm support from the Hong Kong comrades.

Furthermore, 20 doctors and nurses from Hong Kong went to Sichuan to assist the victims and the Hong Kong Hospital Authority sent three specialists to discuss with Beijing's health officials on how infectious diseases in the affected regions could be prevented (*Ming Pao*, May 18, 2008: A7). In fact, many Hong Kong reporters, apart from reporters from mainland China and other parts of the world, flocked to Sichuan province, leading to the exhortation from some earthquake survivors that the Hong Kong reporters should not go to the dangerous regions (*Ming Pao*, May 17, 2008: A25). Some 80 foreign reporters were organized by the State Council to cover the disaster areas in Sichuan, including CNN and Reuters, and they were allowed to report freely on the post-earthquake rescue operation. Reports on the Sichuan earthquake filled the Hong Kong Chinese and English newspapers with pictures showing the disaster regions in detail. One photo on the Hong Kong Chinese newspaper, *The Sun*, showed the face of a man who had been trapped by debris for 73 hours and who painfully waited for rescuers, but he unfortunately died ten minutes after being rescued out from the rubble (*Today Daily News*, May 19, 2008: B7). News reports, website stories and television scenes exposed the tragedies of the Sichuan earthquake to the viewers in an unprecedented manner. The swift response to the Sichuan earthquake came from not only the mainland but also Hong Kong and the international society.

Taiwan's business people in Dongguan responded to the Sichuan earthquake rapidly by holding an urgent meeting and donated 5 million yuan to the rescue operation. The Guangzhou Taiwan businessmen also followed suit and donated 6 million yuan (*Sing Tao Daily*, May 16, 2008: B12). Taiwan's Evergreen Consortium donated US$10 million to the PRC government whereas other Taiwan business people in Shanghai also collected their donations to express their support of the rescue operation. The ROC China Airline airlifted logistical supplies of 110,000 tons to Chengdu; and the Taiwan Red Cross sent 20 professional rescue workers, including engineers, to the Sichuan province on May 16. Taiwan's president-elect Ma Ying-jeou and his wife went to the China Television as volunteers to receive calls from Taiwan citizens who donated money to the Sichuan earthquake's relief work (*Ming Pao*, May 19, 2008: A4).

PERSPECTIVES ON THE ROLE OF NON-GOVERNMENTAL ORGANIZATIONS

Some mainland Chinese scholars have made insightful observations on the role of nongovernmental organizations (NGOs) in the Wenchuan earthquake. Lin Mingang and Zhan Jianhua compared and contrasted the role of NGOs in the Wenchuan earthquake with that in the Taiwan earthquake on September 21, 1999 (Lin and Zhan, 2010: 98–103). Adopting the analytical framework advanced by Chen Jianmin for scholars to comprehend the varying roles of NGOs, government, enterprises, and volunteers in the Taiwan earthquake in September 1999 (see table 2.1), Lin and Zhan have made a number of important observations. First and foremost, the NGOs in the PRC faded away after the Wenchuan earthquake as their sustainability needed the endorsement from the government, while the Taiwan NGOs remained after the earthquake and tended to be more professionally oriented (Lin and Zhan, 2010: 101). Second, Lin and Zhan observed that the Taiwan NGOs became institutionalized from the stage of emergency aid to the phases of resettlement and reconstruction, but the mainland NGOs tended to be noninstitutionalized in the entire process. Third, there was a partnership between the Taiwan government and NGOs in the aftermath of the 1999 earthquake, but the PRC situation in 2008 witnessed the leading role of the mainland Chinese government, which incorporated the role and assistance of NGOs under the rescue and relief blueprint. Lin and Zhan's observations are significant, for they also identified the relatively weaker role of NGOs in the PRC vis-à-vis the counterparts in the ROC on Taiwan. They concluded that "as the relationship between the state and society in current China is 'strong government, weak society,' the patron-client relations [between the PRC government and NGOs] may help to improve NGOs participation in disaster relief" (Lin and Zhan, 2010: 103). In other words, the dependence of NGOs on the government's leadership, blueprint, and directions was inevitable and conducive to the executive-led nature of the rescue and relief operations after the Wenchuan earthquake. The role of NGOs was basically auxiliary while that of the PRC government was of paramount importance.

The research conducted by Lin and Zhan showed concrete examples of the emergence of the dependent type of NGOs in the PRC immediately after the Wenchuan earthquake. For examples, volunteer groups like the Aibai Chengdu Youth Center utilized its own networks on the Internet to send hundreds of information to members and supporters so as to open their own homes to the refugees just one hour after the earthquake (Lin and Zhan, 2010: 99). The Amity Foundation on the night of May 12 quickly collected donations of 1 million yuan, whereas another One Foundation also followed suit. In a survey of NGOs in response to the Sichuan earthquake, 14 percent immediately took action and 73.2 percent did the same within 24 hours.

Table 2.1 The Role of NGOs in Taiwan's Earthquake on September 21, 1999

Stages	*Government*	*Students*	*Enterprisers*	*Volunteers*
Emergency Aid	Leader and implementator	Supporter and information provider	Supporter and resource provider	Assistance and resource integrator
Resettlement	Leader and resource provider	Supervisor and information provider	Provider of resources and services	Coordinator and participator
Reconstruction	Supporter and law formulator	Supervisor and supporter	Provider of resources and services	Coordinator and service executor

Source: Chen Jianmin et al., "Report and Recommendations on Non-Governmental Welfare Organizations Participation in Disaster Relief and Reconstruction in Wenchuan Earthquake," *Social Policy Communications*, vol. 11 (2008).

However, the number of NGOs decreased gradually later; in April 2009 only 50 NGOs remained active in Wenchuan and they involved less than 50,000 volunteers (Lin and Zhan, 2010: 100).

Yet, Lin and Zhan identified the important constraints on NGOs after the Wenchuan earthquake. Under the circumstances that there were very few charity foundations or funds devoted to relieving disasters, the China Children Fund was not sure about whether it could collect donations for the earthquake victims (Lin and Zhan, 2010: 100). The result was the so-called Dujianyan phenomenon, meaning that in Dujiangyan most donations went to a few local charity funds or organizations like the Red Cross, but the crux of the problem is that many citizens wrote checks payable to "Disaster Relief Fund," which was not accepted by banks when charity workers tried to deposit the donations from the members of the public. The situation revealed the absence of legal guidelines and formal relief disaster funds for public donations in time of natural disasters.

Other mainland Chinese scholars have openly argued for the need to engender a scenario in which the PRC government and society would be both "strong" to resist any natural disasters in the future. Zhang Qiang, Lu Qibin, and Zhang Xiulan admitted that the PRC government was basically "strong" and society "weak" from 1949 to 2002, when the Chinese state played a leading role in industrial development, economic modernization, and social control (Zhang, Lu, and Zhang, 2011: 54). They argued that the outbreak of the Severe Acute Respiratory Syndrome (SARS) in 2003 was a turning point prompting the Hu Jintao administration to recognize the need to empower the society so that both the Chinese state and the society could resist the encroachment of any infectious disease (Zhang, Lu, and Zhang, 2011: 54). The 2008 Wenchuan earthquake, according to these mainland scholars, represented another watershed in which the state and society in the PRC were "strong" in their formation and cooperation in response to various crises, including the early 2008 snowfalls and ice pellets that plagued the transportation, roads, highways, and railways in Southern China (Zhang, Lu, and Zhang, 2011: 54). Zhang Lu, and Zhang concluded that a strong society could bring about "social resilience" in the sense that the social "cells" could be nurtured to collaborate with the Chinese state in tackling any crisis and minimizing risks management in the future (Zhang, Lu, and Zhang, 2011: 55).

From a critical perspective, the emergence of a "strong" society accompanying the strong state as advanced by Zhang, Lu, and Zhang was not substantiated. The PRC society remains relatively weaker vis-a-vis the state. Although the Wenchuan earthquake unleashed a series of NGOs involving volunteer groups, they tended to fade away gradually in the process of reconstruction, as Lin and Zhan have accurately portrayed.

Unlike the two perspectives focusing on the dynamic relationships between the state and the NGOs in society, the third perspective articulated by Wu Jianxun and Xu Xiaodi tends to explore the ways in which NGOs interacted with the Chinese state in the Wenchuan earthquake. Wu and Xu have identified four types of interactions between the NGOs and the Chinese state: cooperative, mutual beneficial, contractual, and confrontational (Wu and Xu, 2010: 347–362). They found a variety of nonprofit or nongovernmental organizations emerged immediately after the Wenchuan earthquake, including (1) charity funds like the Red Cross and the Chinese Charity Association; (2) nonpublic charity groups like One Foundation and Nanduo Charity Fund; (3) civilian organizations like Shenzhen Wanke Volunteers Group, Beijing Normal University's Faculty of Social Development and Public Policy, Sun Yat Sen University, and Guizhou Highland Development Study Center; and (4) international organizations such as the Hong Kong Oxfam, Doctors Without Borders, and the International Lions Club (Wu and Xu, 2010: 351). Most importantly, the four types of interactions identified by Wu and Xu showed the dominance of mutually beneficial interactions, like the action of the Beijing Normal University's Faculty of Social Development and Public Policy to mobilize social workers to provide services for earthquake victims, the attempt made by the Lions Club volunteers to build 100 washrooms in Mianyang, and the visits of Greenpeace's personnel to inspect the situation of 100 chemical factories and report their development to the authorities (Wu and Xu, 2010: 359). While Wu and Xu did not mention that those parents of the earthquake victims protesting against the "tofu" buildings could be regarded as a NGO confrontational to the Chinese state, they both emphasized the coordination between NGOs and the state. A minority of groups received "contracts" from the local governments to deal with the relief operation, such as the Nanduo Charity Fund's donation of 10 million yuan for NGOs to continue with their charity services during the reconstruction phase, and the investment of 3 million yuan from Give2Asia and Trafigura to assist the implementation of the disaster relief centers—actions that were sanctioned and legally approved by the PRC government (Wu and Xu, 2010: 359).

In short, three different perspectives on the NGOs persist in the eyes of mainland scholars who studied the post-earthquake developments in Sichuan: the relatively subordinate role of NGOs; the cooperation between the "strong" state and "strong" society; and the partnerships between the state and NGOs.

Recent academic studies of the post-earthquake society of the PRC have founded that cultural products were carefully orchestrated to portray an image of a resilient Chinese society vis-à-vis earthquake and that donations from the private sector were related to the desire of companies to boost their public image and market prices. Florian Schneider and Yih-jye Hwang have found that, in the aftermath of the Wenchuan earthquake, the PRC authorities launched a legitimizing campaign through the utilization of cultural products, including movies, mass media events, the performance-based discourses in the Beijing Olympics, and the Shanghai Expo (Schneider and Hwang, 2013: 636–656). For instance, Feng Xiaogang's movie *Aftershock* was carefully orchestrated so as to portray an image that China was nationally united in the midst of adversities and struggle. Cultural governance was shaped in such a way as to legitimize the PRC party-state.

Some economists who selected a sample of 136 Chinese listed companies found that there was a marked rise in the share prices of those companies which made donations to help the affected regions and victims after the 2008 Wenchuan earthquake (Song, Wang, Yao, and Chen, 2012: 299–315). However, compared with private companies, government-controlled corporations did not have the incentive to donate for the sake of enhancing their reputation. The amount donated by the government-controlled companies tended to be less than those from private corporations. Still, many members of the public were concerned about possible misuse of philanthropic funds, thereby limiting the scale of charitable giving.

THE GOVERNMENT'S REMEDIAL RESPONSES

While NGOs provided the necessary backup for the PRC government to cope with the earthquake disaster, the top Chinese leadership continued to actively take measures to reduce the extent of societal harm caused by the earthquake. On the morning of May 16, President Hu Jintao flew to Sichuan province and joined Premier Wen to inspect the progress of the rescue and relief operation. Hu emphasized that the situation reached a "critical" one demanding extra efforts on the part of rescue operation and the dispatch of an additional 30,000 PLA officers to various regions and villages so as to deal with victims, orphans, and the old and the needy effectively (*Today Daily News*, May 17, 2008: B2). In the afternoon, Hu visited Beichuan and chatted with the rescue workers from the PLA, PAP, fire services, and civilian police. On the night of May 16, Premier Wen held a meeting on the train with the CCH officials, saying that the wreckage caused by the 2008 Sichuan earthquake was "stronger" than the 1976 Tangshan earthquake, and estimating that the death toll would later reach over 50,000 people (see table 2.2). In reality, the actual number of deaths was higher in the Tangshan earthquake than the Sichuan one. Fortunately, the 2008 Sichuan earthquake did not affect the nuclear reactor plants located at Guangyuan and Minyang. In Guangyuan, the nuclear reactor factory was built in 1969 but it was later left idle since 1988, when China joined the nuclear nonproliferation regime. Staff members working in the factory were then laid off, but overall the nuclear reactor plant number 821 remained intact during and after the Sichuan earthquake (*World Journal*, May 17, 2008: B4). Minyang had a nuclear missile plant constructed in 1983, but fortunately it was not affected by the May 12 earthquake in 2008. However, immediately after the earthquake, a team of 21 nuclear security experts were sent to Mianyang and Guangyuan to assess the situation (*World Journal*, May 17, 2008: B4). An emergency nuclear safety plan was triggered so as to prevent any occurrence of a sudden gas leakage of the nuclear power plants in Sichuan. However, in Shifang, a chemical factory had leakage of gas, leading to the immediate dispatch of the PLA's biochemical teams, which used concrete to cover the leakage areas.

The most devastating impact of the Sichuan earthquake was on the schools, many of which collapsed and caused serious death toll. In Beichuan county, 2,447 schools collapsed, whereas Dujiangyan witnessed the complete destruction of 1,800 schools. In response to the tragedy, the Department of Construction in Sichuan province ordered an investigation of why so many schools collapsed during the earthquake. Specifically, if "tofu" buildings were found, the officials and construction companies concerned would likely be penalized. According to the anti-earthquake guidelines of Sichuan's construction authorities, schools that were built should not be over five stories and they were also required to use specific concrete and follow anti-earthquake building standards (*Today Daily News*, May 17, 2008: B3). However, many schools appeared to neglect these guidelines and exposed three main problems: the large and spacious classrooms vulnerable to any earthquake; the relatively huge windows that made the areas of walls relatively small and weak vis-à-vis any shocks; and the construction of external corridors which could be lethal in the event of collapse. Compounding these problems was corruption of government officials, who allowed the use of substandard construction materials in the building of schools and who did not inspect them rigorously and strictly.

Table 2.2 Comparison Between the Wenchuan and Tangshan Earthquakes

	Wenchuan Earthquake	*Tangshan Earthquake*
Time	May 12, 2008 at 14:28	July 28, 1976 at 03:42
Epicenter	Sichuan's Wenchuan	Hebei's Tangshan
Magnitude	7.8-7.9	7.8
Depth	29 kilometers under the surface	22 kilometers under the surface
Time length	Lasting for one minute	Lasting for 15 seconds
Affected regions	Entire China, Thailand, and Vietnam	14 provinces and cities
Affected area	Over 100,000 square kilometers	Over 300,000 square kilometers
Death toll and Injuries	22,069 deaths up to May 17 and 168,699 injuries, but the estimated deaths would be 50,000	242,769 deaths and 164,851 people injured
Nuclear power plants	Minyang and Guangyuan, but the nuclear power plants were not affected	No impact

Sources: *Today Daily News*, May 17, 2008, p. B2; *World Journal*, May 17, 2008, p. B4.

China is one of the most popular countries where Americans adopted orphans. After the Sichuan earthquake, the adoption agencies put off the visits of Americans who visited the province in order to identify and receive their adopted children. The website of the China Adopt Talk received inquiries from Americans on how to adopt the orphans in the Sichuan earthquake. The orphanages in Sichuan's affected cities were in lack of electricity, water, and milk formula, prompting a Chinese American who had been adopted by American parents after the 1976 Tangshan earthquake to lend his helping hand immediately, applying for the adoption of a Chinese orphan (*World Journal*, May 17, 2008: A2).

The PRC government quickly handled the orphans resulting from the earthquake. It decided that children who lost their parents but who passed the university entrance examination would be allowed to study in universities without the need to pay for their tuition fees (*Ming Pao*, May 19, 2008: A5). In Mianyang, classes were resumed in schools so that children would not have their study being disrupted by the earthquake, while the university entrance examination would continue in June 2008. Parents reacted warmly to the government's decisions, saying that as long as their children were alive after the earthquake, it did not matter whether they would be able to enroll in universities for their further studies (*Ming Pao*, May 19, 2008: A5). About

100 volunteers from the China Youth Development Fund played a crucial role in building an "Anti-Earthquake Hope Primary School" for 600 children in Mianyang (*World Journal*, May 19, 2008: A6). The school had nine classrooms and was built in anti-earthquake design. Apart from building this primary school, the fund also collected 100 million yuan to help the earthquake victims in various ways, including the transport of construction materials and tents to Sichuan for the purpose of building more schools for children.

After the first three critical days of the earthquake, the central government in Beijing continued to take the leadership of the entire rescue operation. President Hu Jintao arrived at Wenchuan county where a rock suddenly fell down. Although his lieutenants were concerned about his safety, Hu remarked calmly that the tremor made the houses and buildings unsafe and that safety remained the foremost priority (*Sing Ta Daily*, May 18, 2008: A15). If leadership embraces calmness in dealing with crisis situation, Hu displayed his leadership style in tackling the post-earthquake rescue operation. He added that saving the lives of those who trapped in rubble remained the priority of the rescue agenda, that hygiene condition should be maintained in the affected regions, that necessities such as water and food must be provided to the victims, and that the infrastructure networks should be repaired as soon as possible (*Ming Pao*, May 18, 2008: A21). Finally, he stressed the importance of post-earthquake reconstruction work, although the death toll on the afternoon of May 18 reached 32,477 and the number of injured citizens amounted to 220,109.

Due to the earthquake that altered the landscape of Sichuan, 13 lakes were created by landslides and damaged rivers (*Today Daily News*, May 18, 2008: A2). Nonetheless, these lakes encountered the danger of collapse, leading to floods in the surrounding areas. As such, Qingchuan county had to evacuate 30,000 residents urgently and 3,000 PAP police were sent to help them get out of the dangerous zones to the safer Guangyuan region. According to the assessment of geological experts and the Ministry of Water, some 391 reservoirs in Sichuan were undermined by the earthquake, some showing cracks on the dams, some having landslides, and the shapes of some dams were even distorted. Sichuan province had a total of 6,557 reservoirs, according to Xinhua news reports in 2002. To evacuate 1.1 million residents from danger zones in Qingzhou county, rescue workers and volunteers guided them along the safe routes and ran together toward the destinations (*World Journal*, May 18, 2008: A1). In Beichuan county, 10,000 residents escaped their way out of the dangerous areas where lakes would imminently collapse. The PLA soldiers organized their exit from Beichuan; some carrying the elderly, the injured, and children on their shoulders; and a majority of them reached safe destinations in the afternoon of May 17 (*Ming Pao*, May 18, 2008: A3).

The State Council in Beijing issued an eight-point directive in the rescue operation. First and foremost, the rescue operation had to reach every place without any blind spot (*Ming Pao*, May 18, 2008: A3). Second, medical supplies had to be supplemented and increased so as to save the lives of the survivors and prevent further paralyzed citizens. As a matter of fact, face masks and medicine were deficient in some affected regions, and outdated medicine had to be used in order to tackle health problems (*Sing Tao Daily*, May 19, 2008: B5). Third, within three months, a subsidy of 10 yuan and a certain amount of food had to be dispatched to every victim of the earthquake. Fourth, all the corpses had to be handled properly, together with the human feces, so that infectious diseases would be avoided. Fifth, post-disaster work included the use of crematoriums to deal with dead bodies while orphans and single parents as well as the elderly had to be tackled with great care. Sixth, the monitoring work on post-earthquake tremor had to be consolidated so as to make more accurate prediction and take precautionary measures. Seventh, electricity had to be resumed as soon as possible while preventive work against landslide, collapsed reservoirs, fire accidents, and transport congestion had to be implemented. Eighth, law and order had to be maintained so as to crack down on any theft of logistical supplies and any cheating attempt at squeezing the donations from agencies and citizens. In Dujiangyan, three cadres were dismissed for not making any efforts at rescuing the earthquake victims because they not only responded slowly, but also did their own business even after the earthquake, leading to public outcry and criticisms and eventual purge (*Ming Pao*, May 19, 2008: A17).

On May 18, the PRC government started a three-day commemoration of the dead victims of the earthquake. At 14:28 on the afternoon of May 18, the entire country and its citizens mourned over the earthquake victims for three minutes while the operation of the stock markets in Shanghai and Shenzhen was also terminated for three minutes so as to pay respect to those who lost their lives in the 2008 Sichuan earthquake (*Ming Pao*, May 19, 2008: A1). The PRC national flags in mainland China and overseas embassies as well as consulates were lowered; all official websites in the PRC mourned the dead victims; and entertainment activities and Internet were expected to stop for three days. The transfer of the Olympic torch in the mainland was terminated temporarily for three days. The last time when mainland China formally terminated all its entertainment activities was from September 9 to 18 in 1976, shortly after the late CCP chairman Mao Zedong passed away. Entertainment activities gradually resumed from October 1976 at that time.

During the inception of the mourning days, President Hu and Premier Wen led all the members of the Standing Committee of the CCP Politburo to bow for three minutes. After the solemn mourning ceremony, some citizens who mourned on the Tiananmen Square in Beijing shouted and raised banners saying that China would prevail over any natural disaster, that Sichuan would be victorious, and that Wenchuan would recover quickly.

On the night of May 19, when the Chengdu city government announced that there would be a tremor of magnitude around 6 to 7 in the Wenchuan region, about a million citizens went panic and stayed outdoors in public squares and streets on the morning of May 20. Some built tents outdoors and many were reluctant to stay inside their homes for fear of a large-scale earthquake (*Sing Tao Daily*, May 20, 2008: A1). At 5:00 AM on the morning of May 20, more citizens calmed down and they returned to their homes.

Some mainland scholars have adopted a critical perspective on the role of government in response to the earthquake disaster in 2008. Chen Qiang believed that the governments at all levels demonstrated their responses and performance in several aspects: (1) precautionary detection of the occurrence of the earthquake; (2) the rapid response of government departments; (3) the emergency adoption of the people's principle that regarded the safety and interest of earthquake survivors as of paramount importance; (4) the efficient management of clear division of labor at different levels; (5) the implementation of the tenet of professional treatment to prevent further crises and diseases; and (6) the need for reflections and retrospective evaluation (Chen, 2008: 35–36). However, Chen was critical of the inability of the seismological department and authorities to predict the occurrence of the Sichuan earthquake before May 12, 2008.

As with Chen, Yang Li argued for the PRC's need to improve the system of earthquake prevention and reduction (Yang, 2008: 81–83). He spotted a crucial loophole in the post-earthquake relief, rescue, and reconstruction work, namely a comprehensive blueprint in accordance with the Law on Earthquake Prevention and Disaster Management, especially Article 22, which stated that:

> In accordance with the earthquake prediction and actual situation, the State Council's seismological department and the people's governments above the county level should cooperate in preparing earthquake reduction and prevention plans, which are then approved and implemented by the people's governments. The revision of any earthquake prevention and reduction blueprint should be reported and approved by the departments and authorities concerned (see Appendix 1).

Yang noted that Beijing decided to revise Article 22 and changed it into a new chapter after the Wenchuan earthquake so as to stress the importance of having earthquake reduction and prevention plans at all levels of the government. He commented that any amendment made by the PRC government should take into account several ingredients: (1) the stress on and clarification of the "people's principle;" (2) the categorization of different urban planning zones and features so as to match the information requirements of any post-earthquake relief and rescue operations; (3) the need to listen to public opinion in open consultative sessions so that the earthquake prevention and reduction plans would be more effective; and (4) the necessity of delineating the role of NGOs in the process of rescue, relief, and reconstruction after any earthquake disaster (Yang, 2008: 82). While Chen Qiang bemoaned the failure of the PRC seismological authorities in anticipating the occurrence of the Wenchuan earthquake, Yang called for more comprehensive planning in earthquake prevention and more citizen participation in the formulation of earthquake prevention blueprints.

A MINORITY OF LAW BREAKERS

Although the PRC society on the whole was supportive of the state in the post-earthquake relief operations, a minority of citizens broke the law and tried to benefit from the chaotic situation. Substandard blankets were found by the national quality inspectors after they examined the quality of blankets that were sent to the earthquake victims by some provinces, including Shannxi at which the substandard blankets were originated (*Sing Tao Daily*, May 30, 2008: B5). These blankets were found to have poisonous materials that could and would affect the health of the earthquake survivors using them. The national quality inspectors played a crucial role in checking the health standard of materials and logistical supplies being sent to Sichuan from other places and cities both inside and outside the PRC.

Some selfish citizens violated the law and committed theft and fraud. For example, a few citizens stole the logistical supplies and donations in the affected regions. In Deyang, about 20 rascals stole the tents and food from logistical supply trucks (*Sing Tao* Daily, May 30, 2008: B5). A citizen even disguised himself as a soldier by using a fake soldier identification card in order to acquire logistical supplies like tents in the Chengdu city (*Sing Tao Daily*, May 30, 2008: B5). He was caught for buying the tents at cheap prices and then reselling them for the sake of making profits. Some citizens also worried that the donations from other cities might not reach the earthquake survivors due to bureaucratic corruption. A work edited by two anonymous mainland Chinese criticized the PRC Red Cross for embezzling the tents and charity funds, the Mianzhu and Dujiangyan officials for corruption, some cadres for beating up volunteers, the Chengdu government for overspending in infrastructure projects, and the absence of political accountability of the PRC government to the members of the public over the failure to predict the occurrence of the Wenchuan earthquake (Gong and Wu, 2008).

In response to the possibility of bureaucratic corruption and public maladministration in handling the logistical supplies, the central government in Beijing acted quickly and set up on May 28 a leading small group in supervising and inspecting the anti-earthquake relief and logistical materials (*Today Daily News*, May 30, 2008: B2). The CCP's Central Committee secretary He Yong, who was also the deputy secretary of the Central Discipline Inspection Commission (CDIC) responsible for investigating and controlling the corruption of Party officials and cadres, was appointed as the head of the leading small group. Other members of the group came from the Ministry of Supervision's Anti-Corruption Bureau, the Audit Office that was responsible for checking the expenditure of government departments, the Ministry of Civil Affairs, the Ministry of Finance, and also the CDIC. To control how logistical supplies were distributed and allocated to each affected disaster regions, the central government ordered that a similar body checking the use of logistical materials must be set up at the provincial, county, township and village levels in all the affected places (*Today Daily News*, May 30, 2008: B2).

The highly centralized and hierarchical manner of monitoring the use of logistical supplies can be seen as a hallmark of the PRC's governmental responses to the numerous donations and logistical materials sent from other mainland provinces, cities, Hong Kong, Macao, Taiwan, and other parts of the world. According to the leading small group in Beijing, due to the tremendous support from mainland residents, mainland provinces, overseas Chinese, and also other international agencies, the PRC would have to retain better control over the deployment and utilization of all logistical donations so as to "maintain clean governance in the rescue operation, and to ensure that the materials would really benefit the earthquake survivors" (*Today Daily News*, May 30, 2008: B2). Moreover, any mismanagement, embezzlement, and corruption in the process of dispatching and using the logistical materials would not be tolerated. In response to the call from the central government in Beijing, some 1,700 citizens from the PRC and also outside the country volunteered as the inspectors ensuring good governance in the process of conducting the relief operation. The combination of swift governmental response with volunteer assistance from people inside and outside mainland China stood out as a prominent feature in the politics of post-earthquake relief operation.

The PRC government also paid special attention to the protection of cultural heritage immediately after the earthquake, especially those regions populated with ethnic minorities. Winning the hearts and minds of ethnic minorities was a political necessity in the post-earthquake rescue and relief operations. As such, Premier Wen announced shortly after the earthquake that Beichuan county remained a historical cultural heritage site and hence "we would have to protect the Qiang special autonomous county and cultural heritage" (*Today Daily News*, May 30, 2008: B2). A temporary Qiang special autonomous county administration was quickly established so as to retain the confidence of the ethnic group and to start reconstruction work as soon as possible.

RESPONSE FROM THE TIBETAN ACTIVISTS

During the Wenchuan earthquake, a minority of Tibetan activists stole the logistical supplies in Mianzhu, where 100 of them were holding knives and choppers to threaten the earthquake survivors while grasping the necessities from households (*Ming Pao*, June 1, 2008: A1). In Shifang, for example, a group of Tibetan independence supporters went to households to steal and grasp the logistical supplies, but they were arrested by the police and angry citizens (*Ming Pao*, June 1, 2008: A1). Similar incidents took place in Qinghai where some Tibetan independence activists held weapons to rob the citizens and households of necessities. The majority of Tibetans, however, were law-abiding citizens some of whom were also the victims of the earthquake.

If the Dalai Lama and his supporters are regarded as an interest group, their responses to the Wenchuan earthquake deserve our attention. The summer 2008 Olympics in Beijing aroused a series of anti-PRC government activities inside and outside mainland China, but the Wenchuan earthquake appeared to have a temporary dampening impact on the activities of the Tibetan activists. The Dalai Lama, who was the leader of the Tibetan exile government in Dharmsala, India, expressed his sorrow to the earthquake victims immediately after the earthquake (*Ming Pao Monthly*, June 27, 2008: 28–29). He held a number of Buddhist praying sessions to commemorate those Chinese citizens who died in the earthquake. During his visit to Europe, the Dalai Lama collected donations of 1,500 euros and 50,000 US dollars and tried to give them to the PRC Embassy in the United Kingdom, but the PRC government refused to accept the donations (*Ming Pao Monthly*, June 27, 2008: 29). Although the Dalai Lama made a gesture of goodwill to the PRC authorities, the PRC government remained concerned about his anti-PRC and pro-Tibetan independence supporters. As the Foreign Ministry spokesman Qin Gang remarked: "There will have to be some communications on the PRC government's discussion with the Dalai Lama representatives. At the present, the government departments in the PRC are focusing their time and energies to deal with the post-earthquake relief operations. As such, we hope that the Dalai Lama would create the conditions for discussions and negotiations through concrete actions" (*Ming Pao Monthly*, June 27, 2008: 29). The Dalai Lama and his supporters, who call for Tibetan autonomy and self-determination, have been at loggerheads with the PRC authorities. On March 14, 2008, some Tibetan activists protested against the PRC authorities and sparked a riot that led to confrontation with the local police in Tibet. About 5,000 Tibetans were arrested in the aftermath of the March 14 riots (*Ming Pao* Monthly, June 27: 2008: 29). The May 12 earthquake temporarily dampened the activities of the Tibetan activists.

On May 19, when the Dalai Lama gave a speech in Germany in support of the Tibetan self-autonomy movement, he made remarks critical of the PRC government. The Dalai Lama first mourned the death of the Chinese citizens in the earthquake and added that many parents in the PRC lost their only sons or daughters in the tragedy due to the one-child policy being implemented (*Ming Pao*, May 20, 2008: A22). He appealed to the audience to pray for the dead ones, while insisting that the Tibetans were utilizing peaceful means to strive for self-autonomy. According to the Dalai Lama, "the support of Tibet does not represent any opposition to China" (*Ming Pao*, May 20, 2008: A22). He argued that he was eager to struggle for the "real autonomy" of Tibetan religion and culture, and that he hoped Tibet would become an autonomous entity like Hong Kong where the PRC flag flies. By implication, the Dalai Lama advocated the use of the Hong Kong model to deal with the political future of Tibet.

Indeed, the PRC government was not interested in using the Hong Kong model of "one country, two systems" to deal with Tibet on several grounds. First, the Tibet Autonomous Region, according to the PRC officials, had already enjoyed a high degree of autonomy under the Chinese sovereignty. Unlike Hong Kong and Macao whose sovereignty and administrative right were under the control of the British and Portuguese respectively, Tibet was long a part of the Chinese territory—a position steadfastly held by the PRC government (*Ming Pao,* June 16, 2008: A15). Second, the Dalai Lama, in the eyes of Beijing, was at times inconsistent, sometimes calling for the adoption of self-autonomy in Tibet and constantly criticizing the PRC for its human rights record and policy. Third, and most importantly, the PRC government disliked the attempt of the Dalai Lama to "internationalize" the question of Tibet's future by bringing the Tibet discussions to international forums and countries. Despite the PRC's position, in the spirit of goodwill, the Chinese government representatives continued to talk to the representatives of the Dalai Lama over the question of Tibet. In May 2008, the PRC government requested that both the exile government and Beijing should continue dialogue. The gesture from Beijing was interpreted as a move to demonstrate to the international community that the PRC was keen to discuss with the Dalai Lama side on Tibet's future developments, especially in light of the summer 2008 Olympics that would be held in Beijing (*World Journal*, May 19, 2008: B4). Prior to the Wenchuan earthquake, both sides had already engaged in discussions for six times, although there was no fundamental breakthrough.

Originally, a talk was scheduled in May 2008, but the Sichuan earthquake delayed the process of negotiations between the two sides. The Dalai Lama tried to distance himself from the radical Tibetan elements who triggered the riots in Tibet in March 14, 2008. He remarked that he did not seek Tibetan independence, that he hoped for Tibetan self-autonomy, that he supported the 2008 Olympics to be held in Beijing, that he wished to attend the opening ceremony of the Olympics, and that he desired a situation in which there would be solidarity between the Hans and the Tibetans (*Ming Pao Monthly*, June 27, 2008: 29). The Dalai Lama also grasped the opportunity of the Sichuan earthquake to urge the Tibetans in the Tibetan Autonomous Region (TAR) not to launch any protests because "we should altogether pray for the brothers and sisters of China" (*Ming Pao*, May 20, 2008: A22).

A week after the Wenchuan earthquake and during his trip to the United Kingdom, the Dalai Lama also showed his goodwill intention to the PRC government by outlining four conditions for him to stop his exile life and return to mainland China. He said that, as a refugee staying in India and as an ethnic Tibetan, he hoped that the TAR would be open to foreign mass media, that foreign medical teams and experts would be allowed to visit the TAR, that all political prisoners in the TAR should be released, and that the Tibetan people should exercise their basic human rights (*World Journal*, May 19, 2008: B4). However, his appeals did not bring about the positive response from PRC officials, who tended to be cautious, hard-line and who saw him as still oscillating between a moderate stance and a "radical" position denouncing mainland China on its policy toward Tibet.

Actually, eight days before the Wenchuan earthquake, the PRC representatives contacted the Dalai Lama envoy in Shenzhen, marking the first unofficial dialogue after the March 14 confrontations in the TAR. The talk was initiated by the PRC side, which had been instructed by President Hu Jintao. The PRC government hoped to convey a message that it was "patient and sincere" in fostering dialogue and negotiation with the Dalai Lama side (*Sing Tao Daily*, May 12, 2008: A19). Since the March 2008 riots in the TAR, the Tibetans supportive of their self-determination launched protests in different places, including Nepal and India where protests by Tibetans could be seen every day after March 14. The Nepal police did not allow the Tibetans to stage their protests because the Nepalese government recognized that Tibet was and is a part of the PRC, leading to the clash between the Tibetans and the Nepalese police from mid-March to May 2008.

In June 2008, the Nepalese police arrested 700 Tibetans including three leaders for disrupting public order and engaging in "subversive" activities (*Ming Pao*, June 20, 2008: A23). According to the PRC government, "the Dalai Lama clique" recruited 500 Tibetans in Nepal to participate in protests and activities against Beijing (*Ming Pao*, June 20, 2008: A23). It was reported that Nepal received considerable economic aid from the PRC, whose ambassador in Nepal requested that the Nepalese government should adopt a hard-line approach in tackling the Tibetan protestors (*Ming Pao*, June 20, 2008: A23). In India, 30 Tibetan activists were arrested in June 2008 for trying to cross the border from the PRC in order to disrupt the process of transferring the torch into the Indian territory (*Ming Pao*, June 22, 2008: A21). The politics of anti-PRC protests on the part of Tibetan activists continued nonstop outside mainland China even though the Wenchuan earthquake broke out on May 12, 2008.

Similarly, the anti-PRC activities in Xinjiang province persisted despite the Wenchuan earthquake. Some Uygur separatists demanded the establishment of the Eastern Turkestan state. In late May 2008, they asserted their opposition to the transfer of the Olympic torch in Xinjiang (*Ming Pao*, May 31, 2008: A21). Their spokesman who attended the World Uygur Representative Congress in Sweden criticized Beijing for showing social "harmony" through the delivery and transfer of the Olympic torch, and for continuously "suppressing" the Uygur ethnic group (*Ming Pao*, May 31, 2008L A21). The Wenchuan earthquake had virtually no impact on the Uygur separatists demanding for their own independent state within the PRC.

In June 2008, the Tibetan exile government reiterated the stance of the Dalai Lama, saying that it supported the Beijing Olympics, that it advocated "general autonomy" for Tibet, that it did not call for the creation of any "Greater Tibet" (meaning Tibetans in various Chinese provinces would enjoy self-autonomy and 2.4 million out of 5.4 million Tibetans in the PRC lived in areas outside the TAR, such as Qinghai, Gansu, Yunnan, and Sichuan), and that its spiritual leader was keen to return to mainland China (*Ming Pao*, June 16, 2008: A15). From April 1959 to 1979, the Tibetan exile government and its supporters did call for self-determination and independence for Tibet until 1979, when the late Chinese leader Deng Xiaoping stressed that everything would be negotiable except for Tibet's independence. Since Deng's remarks, the Dalai Lama and his supporters, according to the Tibetan exile government, became pragmatic and emphasized Tibet's autonomy through nonviolent means (*Ming Pao*, June 16, 2008: A15). After the March 14 riots in the TAR, Premier Wen Jiabao appealed to the need for the Dalai Lama to calm down the activities of the Tibetans who confronted the police and who were calling for independence in the TAR. In response to Wen's remark, the Dalai Lama wrote a letter to President Hu Jintao, saying that he was willing to send his representatives to talk to the Tibetan activists in the TAR and appealed to them to resort to non-violent means. Hence, before the May 12 earthquake, there were signs that the PRC government saw the Dalai Lama as a middleman who could and would reduce the potential for violence and confrontation in the TAR. The occurrence of the Wenchuan earthquake served as another catalyst warming the relations between Beijing and the Dalai Lama side, although a breakthrough in their negotiations remained difficult.

Strictly speaking, the pro-Tibetan independence activists who stirred up the March 14 riots in the TAR could be regarded as a "radical" interest group fostering social instability in the PRC. On the contrary, the Dalai Lama and his supporters were seen by the PRC authorities as relatively moderate, thus leading to the appeal from Premier Wen to the possible intervention of the Dalai Lama in the social turbulence in the TAR under Beijing's sovereignty. The Wenchuan earthquake was obviously an event having an unintended consequence of calming down the Tibetan radicals while giving the moderate Dalai Lama a golden opportunity to demonstrate his goodwill to Beijing.

If ethnic minorities were divided into moderates who mourned for the earthquake victims and radicals who continued with their pro-independence activities, political dissidents in the PRC saw the Wenchuan earthquake as a tragedy and tended to tone down all their political activities. On May 28, 2008, a group of the mothers of the students who died in the June 1989 Tiananmen tragedy, made a website in commemoration of their dead sons and daughters (*Ming Pao*, June 1, 2008: A24). During the Tiananmen incident, the PLA was mobilized by the top CCP leaders to clear the Tiananmen Square, where student democrats and workers demanded clean government and reforms in the PRC. The mothers of the dead students, or the so-called Tiananmen mothers, continued to mourn for their loved ones. Their representative, Ding Zilin, said that the nation made progress by lowering the flag to mourn those people who died in the Wenchuan earthquake because this was a show of respect for the dignity of the victims (*Ming Pao*, June 1, 2008: A24). However, she asked: "When will the national flag be lowered for the sake of mourning those who died in the Tiananmen incident" (*Ming Pao*, June 1, 2008: A24)? Ding's son was found to have three bullets on his dead body on the night of June 4, 1989. In order to make more people understand the Tiananmen tragedy, she and other Tiananmen mothers decided to open the website. Ding emphasized that loving a country does not mean the elimination of dissenting views (*Ming Pao*, June 1, 2008: A24). She also expressed her hope that Premier Wen would treat all lives as the same, implying her dream that one day China would mourn the people who died in the Tiananmen tragedy. Liu Xiaobo, another mainland dissident, also expressed his hope that the PRC would one day mourn over those who sacrificed their lives in the 1989 Tiananmen incident. A dissident outside the PRC, Wang Dang, who was one of the leaders of the Tiananmen incident, remarked that the PRC made progress by releasing more news and information to the citizens after the 2008 Wenchuan earthquake, and that "China would really have hope if this information openness can deepen in social life, especially political aspect" (*Ming Pao*, June 1, 2008: A24).

CONCLUSION

The Sichuan earthquake triggered varying responses from the civil society and citizens in the PRC. Unlike the West where civil society is characterized by groups independent of the control of the government, the civil society's growth that could be seen immediately after the Sichuan earthquake is punctuated by pro-government groups, like women, mobilizing their supporters, and volunteers who were far more devoted to rescuing the victims rather than fostering any anti-CCP activities. The responses from other provinces and cities were rapid and warm, pouring logistical supplies, donations, and manpower into the Sichuan province immediately. Although anti-PRC activities continued to be launched by the Tibetan and Xinjiang separatists, these people were of the minority, as with the lawbreakers who attached importance to their own self-interests rather than the collective interest of rescuing the earthquake survivors. The impressive voluntary action on the part of groups and citizens in the PRC in the aftermath of the earthquake was unprecedented, thanks partly to the relatively open release of information by the Chinese government on the progress of the rescue operation through the mass media.

Chapter Three

The Role of the People's Liberation Army

The ways in which the PLA was quickly mobilized to rescue the earthquake victims demonstrated its critical role in responding to crisis circumstances in the PRC. In 2011, it was reported that the PLA leaders were actually unhappy about the fact that Premier Wen Jiabao had criticized the sluggishness of the PLA's initial responses to the Wenchuan earthquake (Hong Kong Cable TV News, September 4, 2011). This chapter explores how the PLA was fully mobilized to tackle the earthquake and its related difficulties in the process of the rescue operations. According to the Regulations on the Army's Participation in Disaster Rescue issued by the Central Military Commission in June 2005, the PLA set up eight crack teams of 50,000 soldiers and officers, each team including specializations in flood reduction, airborne transport support, transport emergency response, maritime search and rescue, standby communications support, medical stabilization and epidemic prevention, and nuclear and biochemical response (*Beijing Review*, September 2, 2010). Although the PLA regulations were provided, the Wenchuan earthquake tested whether the military could manage any natural disaster practically.

THE PLA DEPLOYMENT

On May 19 the Chief Command Headquarters' deputy minister, Ma Jian, openly revealed the detailed mobilization and operation of the PLA. Together with the PLA Chief Command Headquarters, Chief Political Headquarters, Chief Logistical Headquarters, and the Chief Supplies Headquarters, Ma met the mainland Chinese media and admitted that these four headquarters were fully involved in the rescue operation. He further disclosed that 113,000 PLA soldiers went to rescue 20,000 earthquake victims and survivors while dealing with 30,000 injured citizens (*Today Daily News*, May 19, 2008: B3). Ma added that about 200,000 citizens and tourists were resettled and transferred to safe zones. According to Ma, the PLA would not abandon but instead would continue to go into every village so as to rescue the lives of residents. In an unprecedented move, Ma introduced the five main responsibilities zones as outlined in table 3.1 below, showing how the PLA command oversaw the entire operation in the city of Chengdu, where the Chengdu military region's anti-earthquake joint command set up its headquarters. Five regions were created to facilitate the rescue operation: (1) The Wenchuan-Dujiangyan-Yingshou areas where 23,950 PLA officers were sent (including Chengdu military region's brigade of 5,978 soldiers; Jinan military region's armour division of 7,972 officers; and Jinan military region's mechanical and rifle division of 10,000 staff members); (2) the Beichuan-Mianzhu-Mianyang region where the naval and artillery forces of 8,838 soldiers were stationed; (3) the Pengzhou and Chongzhou areas at which the Jinan military region's brigade of 8,550 soldiers and Chengdu's airborne force of 3,242 officers were located; (4) the Shifang-Deyang region where 14,290 officers from the naval and land forces were stationed; and (5) Qingchuan region where 22,862 soldiers from 2 armour and rifle divisions as well as an artillery force of 2,200 soldiers were dispatched. The southern areas surrounding these five disaster regions were stationed with the air force, including about 100 military transport planes and helicopters at Phoenix Mountain, Gong Gorges, and Pengshan near the Chengdu city.

After Ma outlined the five responsibilities regions, he elucidated that the duties and responsibilities of different PLA forces were dynamic and adjusted in accordance with the changing circumstances. A spokesman of the PLA Defence Ministry, Hu Changming, added that the PLA forces were mobilized immediately 14 minutes after the earthquake. Furthermore, the extent of advanced equipments and the degree of PLA mobilization were unprecedented in the history of the PRC's military deployment in response to a noncombat emergency situation. When questioned by reporters whether the nuclear power facilities in Sichuan were affected, Ma delineated that although some military facilities were slightly affected, PLA and PAP forces were deployed to protect the nuclear power plants, which were not really affected by the Sichuan earthquake. Meanwhile, the environmental protection authorities and the Radiation Safety Center were monitoring the development of nuclear power plants immediately after the earthquake in order to prevent any nuclear accident.

Overall, the PLA mobilization was swift and effective. Ten minutes after the earthquake, the PLA set up the Emergency High Command Headquarters to coordinate the mobilization of the air, land, and naval forces in the rescue operation (*Today Daily News*, May 27, 2008: B2). It implemented the orders

Table 3.1 The Stationing of PLA Forces in the Disaster Regions

Responsibilities Regions	*PLA Forces*
Wenchuan-Dujiangyan-Yingshou	23,950 PLA officers were sent, including Chengdu military region's brigade of 5,978 soldiers, Jinan military region's armor division of 7,972 officers, and Jinan military region's mechanical and rifle division of 10,000 staff members
Beichuan-Mianzhu-Mianyang	Naval and artillery forces of 8,838 soldiers were stationed
Pengzhou and Chongzhou	Jinan military region's brigade of 8,550 soldiers and Chengdu's airborne force of 3,242 officers
Shifan-Deyang	14,290 officers from naval and land forces
Qingchuan region	22,862 soldiers from 2 armor and rifle divisions, and an artillery force of 2,200 soldiers

Source: *Today Daily News*, May 19, 2008: B3.

of both the national anti-earthquake relief operation's Chief Command Headquarters and the Central Military Commission. According to the *Liberation Army's Daily*, all kinds of the military were involved in the rescue operation: land, naval, and air forces; artillery forces; PAP forces; civilian and reserves staff officers; commandos; and other special forces such as detection, communications, chemical and engineering, and fire services.

According to a report prepared by the Japanese National Institute for Defence Studies on the PLA activities during the Wenchuan earthquake, the PLA set up the Army Command Group Resisting the Earthquake and Relieving Disaster and it was under the command of the Central Military Commission (*NIDS China Security Report 2012*, 2012: 21). Moreover the Army Command Group supervised the Chengdu military command under which there were other military units. This description from the Japanese research institute was accurate if the Army Command Group actually referred to the PLA Emergency High Command Headquarters, which as mentioned above was set up ten minutes after the earthquake and which was under the dual leadership of the Chief Command Headquarters led by Premier Wen and also the Central Military Commission. However, the Japanese report was inaccurate in one aspect: it depicted the Central Military Commission as being under the direction and command from the CCP Central Committee (*NIDS China Security Report 2012*, 2012: 21). In fact, it was the Standing Committee of the CCP Politburo which was the top body giving the orders from the top Chinese leaders, like President Hu Jintao. Both the Central Military Commission and the State Council's Chief Command Headquarters had to follow the directives or orders from the Standing Committee of the CCP

Politburo. Guo Boxiong, the vice chair of the Central Military Commission, was in fact a member of the State Council's Chief Command Headquarters, or the State Council's Command Headquarters for Resisting Earthquake and Relieving Disaster according to the Japanese report, which did accurately depict the overlapping membership between the two bodies (*NIDS China Security Report 2012*, 2012: 21). Guo's overlapping membership facilitated the coordination between the military and the State Council under the leadership of Premier Wen. The line of command of the PLA during the Wenchuan earthquake is described in figure 3.1.

However, as of May 19, seven PLA soldiers died during the process of the rescue operation. Ma explained that they were unfortunately buried by rubble and debris during the earthquake and its related tremors that led to the collapse of many buildings. Ma also denied the rumors saying that in the process of dispatching paratroopers into the disaster areas, 2 of the soldiers died (*Today Daily News*, May 19, 2008: B3). As a matter of fact, from the CCTV news and other newspapers reports from Hong Kong, the PLA forces tried their very best to rescue the victims, running for miles into the dangerous zones, carrying logistical supplies and the injured as well as the dead on

Standing Committee of the CCP Politburo

I
I
I

Central Military Commission | State Council's Chief Command Headquarters

I
I

PLA Emergency High Command Headquarters

I
I

PLA Military Regions (Chengdu and Jinan)

Figure 3.1. The Line of Command of the PLA in the Wenchuan Earthquake *Note*: The People's Liberation Army has seven military regions: Shenyang, Beijing, Lanzhou, Jinan, Nanjing, Guangzhou and Chengdu. See "PLA military regions" (2002), in http://news.xinhuanet.com/ziliao/2002-01/25/content_254347.htm, access date: June 21, 2014.

their shoulders, removing debris and rubble in the midst of the post-earthquake tremors, and parachuting into difficult terrain in poor weather conditions. The experiences that the PLA forces gained in the 2008 Sichuan earthquake were both unprecedented and invaluable.

On May 25 and 26, about 1,800 commandos carried ammunition and explosives to Tangjiashan in order to blow up a dam in danger of collapse that would bring about flooding in the region (*Sing Tao Daily*, May 26, 2008: B1). Tangjiashan encountered continuous rainfall of between 20 to 30 millimetres, leading to the possibility of the collapse of the dam where a lake had excessive water storage. The Chief Command Headquarters held an emergency meeting in Sichuan and the meeting was presided over by the deputy commander Hui Liangyu, who was a vice premier of the PRC State Council. Hui ordered that the PLA had to seize every minute to blow up the dam so as to release the excessive water from the reservoir. As a result, the PLA deployed the Russian-made MIG-26 heavy transport plane to inspect the situation and touched down in the dangerous area on the morning of May 26. At the same time, the PLA helicopters arrived at the reservoir. Led by the Jinan military region's brigade, the air force reached the Tangjiashan area but the officers did not bring along with them sufficient necessities such as food and water due to the urgent and immediate deployment (*Sing Tao Daily*, May 26, 2008: B1). However, 1,800 commandos made up from the Chengdu military and the PAP started walking into the dangerous areas on the night of May 25, carrying with them more necessities such as water and food together with bags of explosives. Several hours later they arrived at the destination and conducted the emergency work in preparation of blowing up the dam concerned. If the dam suddenly collapsed, it would flood hundreds of thousands of houses and necessitate the evacuation of 640,000 to 720,000 residents in the affected areas. The water release operation was eventually successful without the need to evacuate hundreds of thousands of residents urgently.

The PLA rescue operations included other relief work, including the demolition of dangerous buildings, the collection of lost properties of residents, the assistance in constructing makeshift tents and the help of farmers and residents to collect the wheat from farmland. Many farmers were reluctant to leave their farming plots and land and the PLA had no choice but to assist them in collecting the wheat from their farmland (*Ming Pao*, May 30, 2008: A1). In other scenes, the PLA rescued young children and elderly people, singing songs for those survivors who had birthdays (*Ming Pao*, May 27, 2008: D17). The people's respect for the People's Liberation Army was fully demonstrated in the 2008 Sichuan earthquake's rescue operation because the PLA soldiers made genuine efforts at saving the lives of citizens and conducting relief operations effectively.

THE PLA DIFFICULTIES

The PLA did encounter some difficulties in the process of the rescue operations. Technically speaking, the PLA training and facilities were not conducive to earthquake crisis management. A report in Hong Kong's *Sing Tao Daily* exposed the technical issues surrounding the PLA handling of the earthquake disaster. First and foremost, the mainland-built IL-76 helicopters were relatively large in their size, not only lacking the necessary agility in touching down on difficult and rugged terrains but also showing deficiency in both quantity and quality. With only 60 IL-76 military helicopters and 400 civilian-type helicopters, the PLA aviation production could not cope with the demands of the rescue operations in the 2008 Sichuan earthquake (*Sing Tao Daily*, June 1, 2008: B1). It was hoped that the IL-76 helicopters production would reach 80–100 helicopters in 2010. But the insufficient number of IL-76 was a handicap in the rescue operation. On June 11, a Chinese helicopter, whose type was not named in the Hong Kong news report but appeared to be IL-76, had an accident in Yingshou during the touching-down process because of mechanical problems (*Ming Pao*, June 12, 2008: A18). The helicopter had 13 people and 3 of them were injured, including one serious injury. Another helicopter went to the accident scene to rescue the injured soldiers.

On June 1, 2008, one M-171 helicopter carrying 4 soldiers and 10 injured residents was missing on its way to Yingshou. Upon hearing the news, President Hu expressed his deep concern and urged that the PLA should rescue its own colleagues as soon as possible. The vice chairman of the Central Military Commission in Beijing, Guo Boxiong, went to Sichuan and ordered an immediate rescue of the PLA soldiers after the accident (*World Journal*, June 1, 2008: A1). The helicopters met a sudden air wave amidst poor weather condition and lost contacts with the command headquarters. Eventually, it was found that the number of deaths incurred from the helicopter's accident amounted to 19, including a pilot who was an ethnic minority Qiang (*Ming Pao*, June 22, 2008: A24). However, another report had revealed that 18 people died in the accident of the M-171 helicopter, including five pilots and all the dead bodies were found, analyzed, and sprayed with anti-decay chemicals on the night of June 11 when almost 1,000 PLA, civilian police, and residents went up to a mountain cliff inside a valley and searched the collapsed helicopter (*Ming Pao*, June 12, 2008: A18). The black box recording the flight history and circumstances of the helicopter was located and analyzed later. No matter how many soldiers and citizens died in this accident, the M-171 appeared to demonstrate some technical problems in the midst of the rescue operation, especially under the timid and poor weather conditions surrounding the valley in Yingshou.

Only one Russian-made MN-26T helicopter was available during the 2008 Sichuan earthquake. The MN-26T helicopter was bought from Russia and could transport 120 soldiers or 60 injured persons, while having the capability of carrying 20 tons of logistical supplies (*Sing Tao Daily*, June 1, 2008: B1). In terms of capability, MN-26T was five to six times stronger than ordinary helicopters. Unfortunately, the lack of any additional MN-26T meant that the PLA aerial transport and rescue facilities were relatively weak in the Sichuan earthquake. The PLA only relied on the Black Hawk-17, 8K-helicopter, and other civilian planes in the Wenchuan earthquake's rescue operation.

The Russian-made MN-26T played a crucial role in the rescue operation. It could carry heavy cranes and tractors that removed debris and rubble from collapsed buildings. It was pictured as carrying a 13-ton heavy crane machine into Tangjiashan so as to prepare for the release of excessive water from a reservoir (*Ming Pao*, May 27, 2008: A1). Altogether, the MN-26T transported 15 cranes into the disaster region, although the PLA did toy with the idea of using bombers to blow up the dam where the reservoir with excessive water would run the risk of collapse and flooding the villages and towns nearby. The Wenchuan earthquake changed the course of the original rivers, widening them and filling them with excessive water that could become lethal to many regions. As a result of the explosive work on the dam, excessive water was eventually released and channeled to safer regions without the need to evacuate hundreds of thousands of residents surviving and living in nearby towns and villages.

In terms of communication equipment, the PLA facilities exposed obvious problems. Satellite-controlled telephones and short-wave radios became the crucial communication tools in the post-earthquake rescue operations and communications; however, they were deficient in the immediate aftermath of the earthquake, leading to the phenomenon that the Chinese telegraphic and telecommunication staff members could only bring in the marine satellite facilities for communication one day after the earthquake that happened on May 12. By using satellite-controlled telephones, the PLA rescue team entering the Beichuan county could then communicate with the outside military forces and commands. It was reported that of the 15 paratroopers being sent into dangerous zones, they relied only on a civilian-made satellite telephone to keep their communication line with the command headquarters. Hence, the lack of PLA-produced advanced communication equipments was regarded as a weak point in the modernization of the Chinese military before the 2008 Sichuan earthquake. Although mobile telecommunication stations installed on the PLA vehicles were utilized quickly and one PLA troop used satellite equipment to determine geographical locations on May 13 and 14, they appeared to be rare and insufficient (*Today Daily News*, May 18, 2008: B2). Perhaps the absence of satellite equipments for the PLA to handle the Wenchuan earthquake explained why later the PRC government was determined to accelerate the progress of the space program so that its new satellite facilities would no longer become a hindrance to any earthquake crisis management.

The lack of pilotless drones was seen as another weakness in the modernization of the PLA. Without any unmanned aircraft or drones that could fly into the dangerous zones to detect the seriousness of the earthquake disasters, the PLA could not conduct its rescue operation in the most effective manner (*Sing Tao Daily*, June 1, 2008: B1). The lack of aerial reconnaissance and detection in the Wenchuan earthquake meant that the PLA had to learn a bitter lesson, namely it would be imperative to develop unmanned drones, enhance the production of satellite, and to improve the design and production of helicopters with greater airlifting capabilities of at least ten tons so that the Chinese military would become well equipped in coping with any natural disaster in the future.

During the first critical 72 hours, other technical problems of the PLA and its rescue operations were exposed, including the lack of heavy cranes that could remove debris and rubble, the absence of large mechanical tools that could cut the stones and concrete into smaller pieces, and the deficiency of small-scale equipment that facilitated the entire rescue work (*Sing Tao Daily*, June 1, 2008: B1). Some residents who were trapped in rubble and debris had to wait for at least three days so that the arrival of cutting machines and cranes would be able to remove the rocks and rescue them. Furthermore, sniffing dogs were insufficient as their presence could play a critical role in identifying those residents trapped inside collapsed buildings. The PLA deployed 12 sniffing dogs in dealing with natural disaster for the first time in the PRC history in 2003, when an earthquake broke out in the province of Xinjiang. However, this type of dogs was not the focus of development and hence its inadequacy exposed another loophole in the modernization of the Chinese military.

As a matter of fact, it was an open secret that Premier Wen Jiabao vented his temper at least twice on May 13 and 14 due to the sluggishness of the PLA in the early phase of the rescue operation. When Wen heard on May 13 that the PLA rescue operation was hindered by landslide and bad weather conditions, he yelled at the PLA officer over the phone and then the PLA was quickly mobilized to arrive at Pengzhou areas. On May 14, as Premier Wen heard that 34 of the 58 affected townships and counties did not envisage the arrival of any rescue worker and PLA soldiers, he shouted angrily at a commander of the PLA paratroops and remarked that they needed to understand that the people of China had been feeding them, and that they should go into the disaster areas with great efficiency and effectiveness (*Sing Tao Daily*, June 1, 2008: B1). The mobilization of the PLA in coping with natural disasters in the PRC was not only seen in the 2008 Sichuan earthquake, but also in the 1998 flooding across many parts of China when 300,000 PLA soldiers were deployed, and in the extremely heavy snowfall and icing disasters in Hunan province in January 2008 as 800,000 PLA and PAP forces were sent to resume public order.

Many PLA soldiers were born in the post-1980s and post-1990s era and therefore they were young and inexperienced in dealing with sudden crises, not to mention any war scenario. It was reported that 20 percent of the PLA soldiers being deployed and mobilized to tackle the Sichuan earthquake encountered "psychological shocks," giving rise to health problems such as depression, worries, hallucinations, and even self-injuries (*Today Daily News*, May 30, 2008: B4). When asked by reporters about their feelings, usually the PLA soldiers replied by saying that they had no fear because they were military officers; nevertheless, the reality was different. Even a member of the Japanese rescue team that was sent to Sichuan quitted the rescue operation after he felt guilty of being unable to save the life of a Chinese victim being trapped in debris and rubble (*Today Daily News*, May 30, 2008: B4). If psychological shock and depression marked this Japanese rescue worker, the situation of the PLA soldiers was the same. Some psychological counselors working in the PLA admitted to reporters that the PLA soldiers would need more psychological and spiritual support than conventional wisdom could assume.

As a matter of fact, more PLA soldiers went to the Sichuan hospitals and clinics to receive psychological counseling services. One military doctor in Chengdu told the reporters from Hong Kong's *The Sun* newspaper that, as an example, a PLA soldier could not eat or sleep, and that he remained very quiet (*Today Daily News*, May 30, 2008: B4). As a result, the soldier's superior asked him to visit the psychological counselor in Chengdu city. Post-tremor psychological disorder appeared to be serious and prominent in some PLA soldiers, as with some American soldiers who encountered similar experiences in battles in Iraq and the Gulf War. According to a Chinese PLA psychological doctor, Wang Waimin, if the situation of post-tremor psychological disorder continued for six weeks, the situation of the PLA solider concerned would become very serious. Hence, he and his psychological workers had to grasp the first three weeks to deal with any ailing soldiers showing symptoms of such psychological disorder.

In order to boost the morale of the PLA soldiers, the deputy director of the Chengdu military region's political headquarters, Zhu Xinmin, said 650 sets of television, 170,000 radios, and 14,000 daily military newspapers as well as psychological counselors were sent to the frontline officers in the rescue operations (*Today Daily News*, May 30, 2008: B4). Wang added that his colleagues allowed the young soldiers to release their emotions and sadness during the national commemoration day on May 19, when many soldiers put off their hats and helmets and cried loudly for those citizens who died in the earthquake and post-earthquake tremors. According to Wang, the PLA soldiers imposed a high level of standard on their own work but might feel incapable during natural disasters, and as a result, depression could be the outcome as there were gaps between expectations and performance. Allowing the PLA soldiers to release their emotions and pressures was, according to Chinese military psychologists, a feasible alternative of tackling potential depression and post-tremor psychological disorder.

A military commander told the reporters a story about a young 19-year-old soldier's psychological distress. The young soldier went to a collapsed building to rescue the survivors but was shocked to witness a couple whose bodies in a sofa facing their home television were crushed to pieces by the fallen debris and rubble (*Today Daily News*, May 30, 2008: B4). The 19-year-old man could not sleep for nights and became very anxious and nervous for days until the commander calmed him down. The commander gave him special care in the following days, identifying that he became calmer than before although he was equally active in rescuing the survivors.

Many of the PLA soldiers who were sent to implement the rescue operations were not merely young but also the lone children in their families in the era of the one-child policy in the PRC. Their experiences in the Sichuan rescue operation, according to some military commanders, had the unintended consequence of making them more mature psychologically and in terms of personal development. They had to deal with dead corpses continuously for days and weeks. Understanding the unprecedented situation encountered by the young PLA soldiers, the Central Military Commission decided to send eight psychological teams to the Sichuan province so that more psychological and personal services could be delivered to the young soldiers and officers. On the other hand, actors and actresses were arranged to be sent to the military sites, such as the Hong Kong actor Andy Lau, so that songs and laughs could minimize the pressure mounting on the young soldiers. Lau was warmly welcomed by many PLA soldiers who saw his presence as a great encouragement to their diligent work in the front line (*Today Daily News,* May 30, 2008: B4).

The PLA's psychological work began late in 2007, when the PLA high command realized after the Gulf War that the modernization of the Chinese military would necessitate the development of psychological counseling and services, and as a military exercise in Western China experimented with such services that were warmly embraced by the PLA soldiers (*Today Daily News*, May 30, 2008: B4). The PLA's military science college is now embracing psychological counseling and services delivered to its soldiers and it has also provided training and medical facilities for psychologists to detect and diagnose any psychological problems on the part of the young rank-and-file officers.

THE PRC GOVERNMENT'S DECISION CHANGE REGARDING JAPANESE MILITARY ASSISTANCE

Originally, the PRC government toyed with the idea of requesting the Japanese government that it would consider sending the Self-Defence Force to transport logistical supplies donated from various cities and provinces to the Chinese airports in the nearby disaster regions in Sichuan (*World Journal*, May 29, 2008: A6). The formal Chinese request was made to the Japanese side through the Chinese ambassador to Japan. In response to the Chinese request, the Japanese government decided to send C-130 transport planes to assist the PRC in the rescue and relief operations. The spokesman of the PRC Foreign Ministry, Qin Gang, also remarked on May 29 that if the Japanese Self-Defense Force (SDF) wished to provide its lending hand in the relief operation in Sichuan, the defense ministries from both countries would be able to discuss the matter further (*Today Daily News*, May 30, 2009: B2). He noted that after the earthquake Japan provided lots of logistical support and help, including the dispatch of a rescue team to which the Chinese government expressed its deep gratitude. Qin concluded: "Due to the fact that this disaster situation is very serious, we welcome any country and its military to provide assistance in the transport of logistical supplies" (*Today Daily News*, May 30: B2). When asked whether the PRC would later refuse the Japanese assistance in the event of protests and complaints from the Chinese citizens in their Internet discussions, Qin did not respond but reiterated that the country that would be willing to assist China would bring about negotiation and discussions between the defense ministries concerned.

On May 30, it was reported that the Japanese government decided not to send the SDF to Sichuan; instead, a civilian plane from Japan would transport the logistical supplies to the disaster-plagued province. The Japanese cabinet secretary announced that due to the cautious attitude of some Chinese citizens toward the idea of sending the Japanese SDF to the PRC, the Japanese government decided to dispatch a civilian plane rather than the military to help the PRC (*Ming Pao*, May 30, 2008: A1). A mainland interest group named Civilian Request for Japan to Pay Compensation over the Second World War reacted by saying that, since the Japanese government took into consideration the feelings of most Chinese people and since it did consider humanitarianism, the changed Japanese decision was "appropriate" and beneficial to Sino-Japanese relations (*Ming Pao*, May 30, 2008: A1). Initially, the Japanese government did decide to send the military planes of the SDF in eight flights so as to deliver the logistical supplies to the city of Chengdu; nonetheless, it later altered the decision in light of the complaints and concerns of the mainland Chinese netizens. In response to the Japanese government's decision, the PRC Foreign Ministry reacted diplomatically and expressed its gratitude to the Japanese assistance in any case. The Japanese cabinet secretary explained to the media why the Japanese "changed" its decision; it interpreted the official Chinese request as "asking for the need to help transporting the blanket and tents to the Chinese airports rather than requesting that the SDF conduct activities in the affected disaster regions" (*Ming Pao*, May 30, 2008: A1). The Japanese Defense Ministry also remarked that "it understands and respects the Chinese cultural tradition and heritage, the Chinese national feelings and their national character" (*Ming Pao*, June 1, 2008: A21). In response, the PLA deputy chief advisor Ma Xiaotin expressed his gratitude to the Japanese understanding of the Chinese history and tradition. The militaries on both the PRC and Japanese sides tackled the issue diplomatically and skillfully. While the PRC government was politically sandwiched between its formal request and the feelings of citizens on the Internet, the Japanese government adopted a wise strategy in sending a civilian instead of military plane so that Sino-Japanese relations could be maintained harmoniously without affecting the feelings of some Chinese people.

CONCLUSION

It is an undeniable fact that the PLA was fully mobilized in the rescue operation immediately after the Sichuan earthquake; nevertheless, its deployment displayed some weaknesses, including insufficient helicopters and the psychological shocks and problems on the part of some young military soldiers. Overall, the PLA mobilization and its diligent work in rescuing the survivors won the hearts and minds of the PRC citizens. The people's dimension of the People's Liberation Army was shown as the military soldiers and officers demonstrated their priority over the lives and safety of citizens. As such, the unintended consequence of the Sichuan earthquake was to legitimize the PRC government through the mobilization of the PLA forces. Indeed, the earthquake's rescue operation also exposed the weaknesses of the modernization of the Chinese military, which would have to improve its facilities, satellite-controlled rescue equipment, and communications in any crisis management of another natural disaster in the future.

Chapter Four

Post-Earthquake Reconstruction Work and Policy Learning

The post-earthquake reconstruction in Sichuan province was rapid but not easy, although the donations from other provinces, cities, and countries did lay the foundation of rapid reconstruction work. This chapter examines how the PRC government dealt with the post-reconstruction work in the Sichuan earthquake after May 12, 2008 and how it learned from the disaster management in any following earthquake. Immediately after the earthquake, it was calculated that the economic loss incurred would amount to US$200 million with a considerable downward pressure of 0.2 percent on China's GDP growth in 2008 and another 0.3 percent decrease in the PRC's industrial production, apart from the inevitable reduction in agricultural production (*Zhonghua Tangshou*, May 30, 2008: 38–39). Despite the gloomy forecasts made by economists on the PRC, the Chinese leaders were resilient and persistent in the post-earthquake reconstruction work, hoping to bring about relatively speedy recovery to the Sichuan province. As early as May 23, the central government in Beijing set up a post-earthquake reconstruction planning group headed by the State Development and Reform Commission and its members included representatives from the Sichuan provincial government and its housing and reconstruction agencies, and also officials from the State Council and other provinces such as Shannxi and Gansu (Wang, 2013: 197).

CENTRAL PLANNING AND PROVINCIAL ASSISTANCE

In light of the earthquake tragedy and the subsequent financial losses of both victims and banks, the PRC monetary authorities in late May issued a directive on how to handle banks mortgage and loans. It said that all the loans, mortgages, and insurance as well as compensations would be annulled, including Mianzhu's 20 billion yuan loans from banks (*Today Daily News*, June 1, 2008: B2). Over 80 percent of the buildings in Mianzhu were regarded as dangerous and they had to be demolished by using explosives. Overall, banks, both commercial and state-owned ones, suffered a total loss of about 450 to 500 billion yuan due to the Wenchuan earthquake, while banks that listed their stocks in the stocks market lost 180 to 200 billion yuan, thus reducing at least 3 percent of the profits of all the banks in 2008 (*Today Daily News*, June 1, 2008: B2). The central government's monetary directive was aimed at settling the financial loss of both banks and citizens once and for all and then focused on reconstruction work at once.

In June 2008, the PRC government published a document, entitled "Regulations on the Post-Earthquake Reconstruction of Wenchuan," outlining the blueprint assisting Sichuan's recovery and reconstruction. The regulations specified the need for other cities and provinces to assist all the disaster-affected regions effectively, emphasized the partnership between the government and societal participation, stressed the importance of reviving the production and self-renewal of the disaster-affected zones, and reiterated the significance of combining the principle of socio-economic with ecological development, adopting the tenet of merging quality with quantity, and persisting with the notion of reconstruction through the help from a partnering province or city (Wang, 2013: 191).

Beijing paired up Sichuan with Guangdong, with the latter assisting the former for three years (*Today Daily News*, June 23, 2008: B5). Moreover, Shenzhen would be expected to assist the earthquake-plagued regions in Gansu province. The blueprint prepared by the State Council aimed at linking one province with another, and one mainland city with another that was undermined by the Wenchuan earthquake. Altogether 18 provinces and cities were mobilized by the central government in Beijing to help the recovery and reconstruction of Sichuan and other affected regions. These provinces and cities included Guangdong, Jiangsu, Shanghai, Zhejiang, Beijing, Liaoning, Henan, Hebei, Shanxi, Fujian, Hunan, Hubei, Anhui, Tianjin, Heilongjiang, Jiangxi, and Jilin. Chongqing city was responsible for the recovery and reconstruction of the western regions of Sichuan province.

Guangdong province was expected to contribute one percent of its local financial incomes to assist Sichuan's Wenchuan county continuously for three years. This meant that of the 2,785 billion yuan of local revenues in Guangdong in 2007, the provincial government would have to invest at least 27 billion yuan in Wenchuan (*Today Daily News*, June 23, 2008: B5). Guangdong also divided the reconstruction phase in Wenchuan into two stages: the period from 2008 to 2010, and from 2010 to 2020. In the first stage, reconstruction work would focus on making 30,000 a community in which there would be housing units of standardized sizes to accommodate urban families and farming households. The planning design of these housing units would embrace markets, schools, kindergartens, and medical service centers. The second stage would expand reconstruction work to accommodate a population of 50,000 in a community. The hidden idea was to adopt the Guangdong mode of reconstructing the Wenchuan county, making the latter recover quickly and follow the success model of the Guangdong province. Hence, the central government in Beijing skillfully exercised its power to instruct various provinces and cities to assist Sichuan and other affected local regions, like Wenchuan, in a strategic and top-down manner. The dynamics of central-local relations could be seen after the Wenchuan earthquake as Beijing used its authority to designate provinces to reallocate their resources to assist the affected regions plagued by the 2008 earthquake.

According to the plan of the central government, each of the twenty-one provinces and cities designated by Beijing would have to help one place disrupted by the Wenchuan earthquake. For instance, Hunan province had to help Pengzhou city, Heilongjiang responsible for Wenjiang precinct, Jiangsu province assisting Mianzhu city, Liaoning helping Anhui province, Fujian dealing with Li county, Shanghai assisting Dujiangyan, and Zhejiang province providing its helping hand to Qingchuan county (*Today Daily News*, June 23, 2008: B3). To help Qingchuan county's reconstruction, Shanghai sent workers and officials from its water department and construction department to repair infrastructure projects in Mianzhu and Shifang. About 3,000 construction workers and officials from Shanghai arrived at the destinations to build 25,000 housing units prior to the deadline on May 25. The construction workers from Shanghai also set up 74 water supply stations in Qingchuan county, providing water to at least 140,000 population continuously for at least two years. The Shanghai officials also set up a coordination office to achieve efficiency and effectiveness in the recovery and reconstruction of Mianzhu and Shifang. Similarly, Hubei province sent workers to arrive at Hanyuan county and they constructed 600 housing units every day, while trying their best to achieve the target of having 10,000 housing units ready before June 25, 2008. The full-scale mobilization of mainland provinces and cities to help their counterparts, which were damaged by the Wenchuan earthquake, was not only unprecedented but also illustrative of the central government's planning in the process of reconstruction.

Beijing itself took the leadership in the entire process of reconstruction. It sent a special team in late May to find out the needs of Shifang city. Then in early June it set up a leading small group to discuss the measures that should be adopted to assist the sister city. Beijing also planned to assist Shifang continuously for three years, while reallocating 10 percent of its financial revenues in 2008 to help Shifang (*Today Daily News*, June 23, 2008: B3). The leading small group in Beijing anticipated that it would take 15 rather than 20 years for Shifang to fully recover from the Wenchuan earthquake. It galvanized various departments together to deal with Shifang's industrial production, including the department of civil affairs, the transport department, the education department, the department of health, the agricultural department, and the department of housing. Moreover, the Beijing audit office sent 150 officials to Shifang to audit and monitor the use of logistical supplies, donations, and expenditure, adopting central control of local use of resources. The Red Cross and other charity organizations were also encouraged to supervise the allocation of logistical supplies strictly and cautiously. Finally, Beijing helped Shifang build a social welfare center that would accommodate the poor and the elderly, the orphans and also the disabled (*Today Daily News*, June 23, 2008: B3). The entire implementation of Beijing's planning for Shifang's reconstruction was conducted in a top-down manner, ensuring good governance and avoiding any public maladministration at the local level.

After hearing that the central government in Beijing assigned different provinces and cities to assist the disrupted regions in Sichuan, the Hong Kong government delegation led by Chief Executive Donald Tsang went to Sichuan for a three-day visit in late June 2008, also trying to assist a particular region in the process of reconstruction (*Today Daily News*, June 24, 2008: C4). The Hong Kong government wished to cooperate with the business people, especially in the Hong Kong construction and engineering sectors, so that any assistance in reconstructing Sichuan would become collective and more effective.

On August 27, 2008, the central government in Beijing held the State Council's meeting which approved a blueprint for reconstruction after the Wenchuan earthquake. The blueprint emphasized the need for better planning and integration in several aspects: rural-urban population and industrial production patterns, the rural-urban characteristics in construction and consumption patterns, the need for constructing new buildings that would be resistant to earthquake, the importance of dealing with rural poverty, the better promotion of education and social welfare facilities, the care for ecological protection and the safety of energy supplies, the mobilization of enterprises in the process of reconstruction, and the emphasis on "the excellence of the Chinese cultural tradition and the anti-earthquake spirit" (Wang, 2013: 198–199). Obviously, the CCP leaders hoped to have a macro-level comprehensive and integrative plan that would galvanize the support of the public in the reconstruction process after the Wenchuan earthquake. The blueprint served an important political function of legitimizing both the Chinese state and the CCP in the process of reconstruction.

The spirit of anti-earthquake ability was publicized by the official media. In Qingchuan, for example, banners were raised to encourage residents to use their hands and legs to win over any natural disaster (Wang, 2013: 203). This so-called Qingchuan spirit was stressed by the local villagers and Premier Wen's visit to the region in August 2008 reiterated the same theme, saying that the residents would have to demonstrate their struggling spirit so as to retain self-sufficiency and achieve speedy recovery. Stories about how villagers who lost their family members and properties could return to normalcy through persistent struggle and perseverance abounded, thus merging local heroic stories with the central government's intention of utilizing the reconstruction phase to legitimize the Party-state.

COLLECTION OF DONATIONS AND CRITICISMS OF THE NEW RICH

Reconstruction would not be possible without the collection, distribution, and allocation of donations from other provinces, cities and countries in the first place. In aftermath of the Wenchuan earthquake, donations continued to pour into the Sichuan province. In Holland, the overseas Chinese groups and individuals, including the Europe Hangzhou Joint Association and Chinese community leaders, donated 500,000 euros and submitted their donations collectively to the Chinese Embassy in the Netherlands (*Sing Tao Daily*, May 21, 2008: A19). In Hong Kong, business tycoon Li Shau-kei's eldest son, Li Ka-kit, went to Sichuan province together with 80 social workers to assist the earthquake survivors, but before they reached Sichuan Li Ka-kit brought along HK$4.6 billion to both the Liaison Office in Hong Kong and Beijing's officials on how the monies would be allocated to the affected regions (*Today Daily News*, May 26, 2008: C3). Li contacted a professor at Tsinghua university to train the social workers from Hong Kong and he also contacted Putonghua-speaking psychological counselors in Taiwan and Singapore so that all the volunteers would galvanize their strength to assist the Sichuan earthquake survivors.

In Canada, a group called GlobalMedic responded to the Sichuan earthquake by contacting the PRC officials in Toronto for permission to offer help because it included paramedics, police, and firefighters who would like to volunteer to go to Sichuan and teach people how to make water drinkable (*Toronto Star*, May 26, 2008: A4). GlobalMedic director Rahul Singh told the media that a two-person team in the PRC carried with them 10 portable purification units, each of which could supply drinking water for 1,000 people. The team also took along one million donated Irish-made purification tablets, while the GlobalMedic also made air shipment of more purification units that could supply water to 70,000 people in Sichuan. These units were also used in Sri Lanka after the 2004 tsunami.

Despite the continuous donations from people and places outside China, some mainland citizens were critical of the new rich in the PRC, saying that the rich people in the mainland did not react to the Sichuan earthquake warmly. A rich capitalist named Wang Shi's land developing company Wanke only donated two million yuan, leading to public criticisms and his open apology. Wang eventually donated one billion yuan to the Sichuan earthquake survivors and victims (*Sing Tao Daily*, May 30, 2008: B3). The Alibaba's chief executive officer Ma Yun was originally rumored by false news on various websites that he merely donated one yuan, but he quickly dispelled the unfounded rumor by donating one million in person in addition to 47 million yuan from his company. Some citizens discussed on the Internet and concurred that those people who had strong ability should shoulder more responsibility for donations to the Sichuan earthquake's relief operations. They raised the critical issue of the corporate social responsibility on the part of mainland Chinese enterprises. In a public opinion poll conducted by the *Chinese Youth Daily*, 75.9 percent of the respondents replied that they did make donations once every five years in the past; 73.5 percent thought that the image of the rich people was tarnished because of their meagre donations; and 91.7 percent said that they had already donated monies to the Sichuan earthquake victims (*Sing Tao Daily*, May 30, 2008: B3).

Indeed, while state owned enterprises in the PRC are often under the obligations to donate monies to the government during natural disasters, other private enterprises saw donations as lacking the necessary institutionalization and legal parameters, including the issue of how charity organizations were legally defined and managed (*Sing Tao Daily*, My 30, 2008: B3). Hence, the Wenchuan earthquake and its subsequent donations from the public and private sectors raised a whole range of unresolved issues, such as the implementation of the concept of corporate social responsibility among the enterprises, both private and public, and the question of whether charity organizations should be defined clearly and institutionalized.

As a minority of lawbreakers grasped the golden opportunity of reconstruction to benefit themselves at the expense of the interest of the earthquake victims and survivors, the local police monitored the allocation and use of logistical supplies closely. For instance, in Beichuan county, the police observed that, once the residents were allowed to return to their homes in towns and villages to collect their belongings, some outsiders sneaked into the regions, stole the properties of others, and accumulated the stolen materials in housing units (*Ming Pao*, June 28, 2008: A24). As a result, the local police registered all the suspicious logistical supplies and properties of the households, serving as a deterrent against those lawbreakers who stole others' private properties. Other examples of lawbreakers cheating the monies from donations were numerous. In Jiangxi province, four criminal elements used fraudulent phone messages to cheat citizens to donate money in the name of helping the Sichuan earthquake's survivors, but they were identified and arrested by the local police (*Sing Tao Daily*, June 16, 2008: B6). The prerequisite of reconstruction was to maintain law and order, highlighting the important role of the local public security bureau.

ACCELERATION OF THE CARE FOR THE "THREE ISOLATED" GROUPS

During the tenth meeting of the Chief Command Headquarters presided over by Premier Wen on May 19, the need for accelerating the adoption of orphans and the care of the disabled and the elderly after the earthquake was discussed. The Ministry of Finance also targeted at the so-called three types of isolated citizens: orphans, elderly, and the disabled. It ensured that each of them would receive 600 yuan of subsidy from the central government, while social groups and organizations at the provincial and local levels were mobilized to care for the three isolated groups. If the number of "three isolated groups" increased, Beijing's plan was to resettle them to reside in areas surrounding the Sichuan province, such as Gansu and Ningxia. The Ministry of Civil Affairs received daily phone calls from citizens on how they could and would adopt the orphans resulting from the earthquake disaster. The central government also made a series of measures to tackle the orphans: looking for their possible parents and relatives who might be alive, resettling them to reside in welfare organizations and schools that delivered better services, and mobilizing the society to adopt them through educating the potential adopted parents on how they could and should deal with the orphans (Wang, 2013: 180). In reality, the "three isolated groups" benefited from the central government's immediate relief measures, especially the monthly subsidy. In August, an elderly businessman was delighted to use the government subsidy of 8,000 yuan to build his temporary home by collecting old bricks and buying the necessary materials (Wang, 2013: 181). Moreover, the government provided him with a nominal rent of 180 yuan per month so as to help him to recover from his loss of income. On the whole, the government took quick measures to address the needs of the poor and the needy—a crucial strategy that could win the hearts and minds of the people.

RESETTLEMENT

Resettlement of earthquake survivors was a crucial task of the PRC government in the post-earthquake reconstruction work. Beijing took the lead in having the Standing Committee of the National People's Congress (SCNPC) approve a revised Law Preventing Earthquake and Reducing Disasters in the PRC in December 2008 (Wang, 2013: 192; for the content of the original law see appendix 1 of this book). The law focused on how to plan ahead in order to avoid another earthquake disaster, how to predict the occurrence of earthquake, how to conduct rescue operations, and most importantly how to deal with the resettlement of earthquake survivors and victims. The newly revised law stressed the need for a new place to be found to rebuild and resettle the refugees resulting from any earthquake, the importance of reaching stringent anti-earthquake criteria in reconstruction work, and the imperative of avoiding earthquake-prone zones in the process of reconstruction and building new homes for the earthquake survivors.

Resettlement work began immediately in the affected regions of the Wenchuan earthquake. In Mianyang, 61 resettlement points were established and 1000 tents were set up for the earthquake survivors to restart their lives normally (*World Journal*, May 30, 2008: B1). Over a million residents in Dujiangyan were resettled and arranged to live in new homes as early as on May 26, 2008. A first batch of 1,500 Dujiangyan residents went to reside in special housing units (*Today Daily News*, May 26, 2008: B3). There were a water supply station, a washroom, and also a garbage collection point for every 50 housing units. For every 1,000 housing units, there would be a small primary school, a medical clinic, and also a rice and food stall. For every 2,000 housing units, there was a secondary school. This setup was designed to give accommodation to the earthquake survivors in an orderly and systematic manner, allowing them to have some sense of community belonging and letting them recover from psychological shocks more easily and quickly. Most earthquake survivors enjoyed their stay and rest in the special housing units. Although post-earthquake tremors persisted, the earthquake survivors residing in the special housing units were not frightened.

JOBS AND RECRUITMENT

In late May 2008, Chengdu city organized a large-scale recruitment drive aimed at resuming jobs for the victims and survivors of the earthquake. The theme of the Chengdu recruitment was to "promote employment and encountering difficulties," followed by various recruitment drives by other cities in Sichuan from early to late June (Wang, 2013: 177). In Guangyuan city, for instance, the recruitment drive provided 3,000 jobs for the residents affected by the Sichuan earthquake. In Dujiangyan, Chongzhou, and Pengzhou cities, enterprises provided a total of 41,699 jobs for the residents. Guangzhou's 40 enterprises went to Wenchuan and recruited 3,000 residents. The provision of job opportunities and the resumption of work became the morale boosters assisting the victims and survivors of the Wenchuan earthquake.

In the process of reconstruction and rebuilding homes for the refugees, the PRC government learned a lesson from the Wenchuan earthquake and paid special attention on the need to draw up "functional zones" in September 2008 (Wang, 2013: 193). By functional zones, they referred to those districts which should be ecologically protected, like cultural heritage sites and rare forest with special species, and those districts which were suitable for settlement as well as those appropriate for farmland. In other words, proper land use and planning were on top of the priority of reconstruction and redevelopment in Sichuan province. The objectives were to ensure that "families would have their own homes, households would have their jobs, everybody would have protection, facilities would be enhanced, economy would be developed, and ecology would be improved" (Wang, 2013: 195).

INFRASTRUCTURE AND INDUSTRIAL DEVELOPMENT

The Sichuan provincial government was under criticism in late May because its first infrastructure project was to spend 14 million yuan to build a railway linking Chengdu city with the devastated Dujiangyan. The provincial government of Sichuan signed a formal agreement in late May 2008 with the Ministry of Railway to construct this railroad, with the intention of assisting the reconstruction of the earthquake-affected regions in Dujiangyan quickly (*Ming Pao*, May 30, 2008: A21). However, many citizens complained on the Internet that the huge amount of investment should better be spent in more reconstruction work in different regions.

The resumption of industrial development was one of the priorities of the reconstruction process. Sichuan's Dongfang Vehicle and Mechanical Company, which was set up in 1965, played a crucial role in demonstrating how an enterprise devastated by the earthquake could recover rapidly. After the earthquake, the company established an anti-earthquake relief headquarters, mobilizing 2,000 staff members and cadres to form a self-aid team, rescuing 1,000 injured colleagues and also 5,000 other citizens, including the elderly, children, and disabled (Wang, 2013: 212). Premier Wen came to greet the success of the company, which became a model of emulation by other enterprises. In September 2008, it was reported that the Dongfang company resumed production of electricity appliances and vehicles although it had suffered a financial loss of 10 billion yuan. The rapid recovery of the Dongfang company provided a perfect model for the CCP to publicize how the state-owned enterprise was successful in its revival and reproduction.

ANTI-CORRUPTION AND PUBLIC MALADMINISTRATION

The collapse of many "tofu" buildings, especially schools, killed a considerable number of citizens and school children. As such, the local governments in Sichuan, including Beichuan and Mianyang where "tofu" schools caused the grief and dissatisfaction of many parents, decided to look into the quality of the construction of these schools (*Today Daily News*, May 26, 2008: B3).

In Mianyang, for example, local officials first retained the collapse sites and set up an investigatory committee that invited the participation of parents whose children died in the earthquake. Construction experts from Tsinghua University, the Ministry of Education, and the Department of Housing and Rural-Urban Construction were invited to share their views in the committee meetings. While the Beichuan county government put forward a blueprint to plan for the reconstruction of 38 schools that would strengthen their resistance to earthquake of a magnitude 8, the Mianyang officials were also keen to ensure that the reconstructed schools would meet stringent criteria of construction that would counter a similar earthquake in the long run.

The Central Discipline Inspection Commission (CDIC) and the Ministry of Supervision at the central government level issued clear directives in late May to forbid bureaucratic corruption, which would impede the process of rapid reconstruction. The CDIC published a directive on penalties on those who violated the law and regulations in managing anti-earthquake relief logistical goods. Those misbehaviors included the following aspects: using the name of rescue and relief operations to cheat and deceive people, illegally collecting donations from citizens, squeezing the donations from the public, asking people to pay for illegal fees, delaying the process of distributing logistical supplies to citizens, and using fake names to collect logistical supplies (*World Journal*, May 30, 2008: B1). For those officials who violated these rules, they would be demoted, dismissed, and penalized. For those who squeezed the logistical supplies from citizens, their activities would be construed as "corrupt." Moreover, the Audit Office also sent officials to the PRC Red Cross to ensure that there would not be any maladministration in handling its donations and expenditure. At the same time, the Ministry of Supervision and the Ministry of Civil Affairs coordinated among themselves to make sure that the logistical supplies would be handled properly without any maladministration or corruption.

For officials who were found to be negligent, they were criticized and even dismissed in the reconstruction phase. It was reported that President Hu Jintao refused to get off the plane when he saw local officials laying down a red carpet upon his arrival at a disaster zone. Hu said angrily: "How can these officials behave like this? How can they do things for the sake of the people's interest" (*Sing Tao Daily*, June 1, 2008: B4)? The official Xinhua news agency published an article critical of the misbehavior of local officials, criticizing them as having low standard of governance.

Although the anti-corruption campaign has become prominent since the Xi Jinping presidency in the PRC in 2013, there were reports revealing the dark side of the Wenchuan earthquake. Businessman Liu Hong, who donated monies to a school in Beichuan unaffected by the earthquake at all, asserted to the foreign media that he had actually utilized the local criminal gangsters to threaten another business rival who tried to bid for a piece of land in

Beichuan (*Apple Daily*, Febraury 21, 2014: A31). He even relied on the supporters of the corrupt public security chief Zhou Yongkan, who was arrested in early 2014 for his corrupt activities, to persecute the business rival.

Some corrupt officials in Sichuan did face penalties, however. The deputy mayor of Guangyuan city, Wu Lianqi, was investigated twice after the Wenchuan earthquake as he was criticized for embezzling the logistical supplies (*Sing Tao Daily*, Februrary 21, 2014: A24). Eventually, he was dismissed from the CCP in 2014 in the wake of another charge of corruption and embezzlement. In May 2013, the PRC Red Cross publicly apologized to citizens for being criticized as squeezing some of the donations made by mainland artists (*Hong Kong Daily News*, May 1, 2013: A10).

The role of netizens in exposing any corrupt behavior of government officials has continued to be significant since the 2008 Wenchuan earthquake. In June 2012, for example, three Mianzhu officials, including village and township ones, were criticized by netizens for corruption and exploitation of the logistical supplies after the Wenchuan earthquake (*Hong Kong Economic Times*, June 15, 2012: A25). Although netizens might not be able to force the local governments concerned to address the problem of bureaucratic corruption effectively, their daring criticisms helped the central government in Beijing and provincial authorities to identify and investigate possible maladministration and official graft.

In April 2013, Sichuan's Yaan city had an earthquake of magnitude 7 killing 200 residents and its party-secretary was quickly dismissed from the CCP for not only corruption but also having mistresses (*Sing Tao Daily*, November 18, 2013: A28). His misdeeds were highlighted by citizens on the Internet, including a possible embezzlement of some of the HK$1 billion donations from Hong Kong. Although the city government denied that the party-secretary committed any mistakes, the organization department of the Sichuan provincial government looked into the accusations and eventually fired him. Under the circumstances in which clean government has been stressed in the Xi Jinping era, local officials found to be corrupt encounter immediate penalties, especially in areas devastated by any earthquake.

CORRUPTION AND POLITICS OF ANTI-DONATIONS IN HONG KONG

Corruption in the PRC gave rise to the politics of anti-donation in Hong Kong, where some pro-democracy politicians refused to approve the Hong Kong government's attempt at seeking the Legislative Council's approval of its donations after the outbreak of the Wenchuan earthquake in the mainland. Public and private sector corruption in the PRC was a concern often highlighted by the Hong Kong mass media. In June 2011, a girl named Guo Meimei who claimed to be a high-ranking executive in the Red Cross Society of China showed off her luxurious lifestyle in the social media, prompting public criticisms (Carter, 2013). Although it was later found that she did not work for the Red Cross, the incident tarnished the image of the charity organization especially as it represents a crucial channel collecting donations from the public and private sectors.

After the Wenchuan earthquake, the Finance Committee of the Hong Kong Legislative Council in May 2008 approved a provision of HK$350 million for injection into a disaster relief fund, which transferred the donation to the PRC State Council Headquarters of Resisting the Earthquake and Providing Disaster Relief, and to other organizations like the Hong Kong Red Cross, the Amity Foundation, Cedar Fund, Oxfam Hong Kong, and World Vision Hong Kong (Constitutional and Mainland Affairs Bureau, July 2008: 14). From July 2008 to July 2009, the Legislative Council's Finance Committee again approved HK$9 billion for the trust fund in support of reconstruction work in Sichuan (Constitutional and Mainland Affairs Bureau, and Development Bureau, December 2011: 2). Embracing the HK$1 billion donated by the Hong Kong Jockey Club and members of the public, Hong Kong actually raised a total of HK$10 billion in support of Sichuan's reconstruction.

In 2013, some Hong Kong democrats raised a serious question of whether the donations from the Hong Kong government to the relief measure of the mainland's earthquake were really worthwhile. Raymond Wong Yuk-man severely criticized the Hong Kong government for misusing public money due to public mismanagement and corruption in mainland China (Wong, 2013). Wong argued persuasively that the Hong Kong government expressed its political "loyalty" to the central government through the HK$100 million donations after the Yaan earthquake of a magnitude 7 took place in Sichuan in April 2013 and killed 200 people and injured 11,000 residents (*The Standard*, April 22, 2013: 1). Nevertheless, there was absolutely no "oversight" on how the money was spent in the mainland, including the building of social welfare homes in Gansu province where the Wenchuan earthquake affected some mainland residents. Wong summed up the public sentiment by emphasizing that many Hong Kong people did not trust Beijing and the Hong Kong government. Another legislator, Chan Wai-yip, who was a member of the People Power, threatened to deploy filibustering tactics to obstruct the government's efforts to seek the approval of the Legislative Council to make donations to the PRC after the Yaan earthquake. He remarked that "we fear the donations will not be utilized fully under the lack of transparency and monitoring" (*The Standard*, April 24, 2013: 1). Angered by the open criticisms of Wong and other democrats, the official media, *People's Daily*, argued that the "anti-donation campaign" by some Hong Kong people aimed at "politicizing" the rescue and relief operations in the mainland, and that they should not regard the PRC's economic affluence as an excuse of rejecting donations to help the earthquake victims (*Sing Pao*, May 1, 2013: A14). In any case, public mismanagement and bureaucratic corruption in the mainland's handling of donations from Hong Kong triggered fierce debates in the Hong Kong Legislative Council in 2013 over whether such donations were a waste of public money.

The official handling of relief work after the Yaan earthquake did bring about public criticisms of corruption in the PRC. Citizens on the Internet and blogs questioned whether the Red Cross embezzled 80 million yuan of donations, but the official mouthpiece *People's Daily* defended the Red Cross (*Hong Kong Economic Times*, May 1, 2013: A10). Even worse, some parents complained in public that their children who failed to seek a minimum amount of donations were punished, while 100 artists who had their paintings auctioned and gave the proceeds of 84 million yuan to the Red Cross for the victims of the Wenchuan earthquake also questioned where the donations went (*Hong Kong Economic Times*, May 1, 2013: A10; *Hong Kong Economic Journal*, May 1, 2013: A12). In response to public criticisms, the Red Cross apologized to the artists, clarified that the donations went to the earthquake victims for rebuilding their homes, and it openly announced that there was miscommunication with the donors. However, the credibility of the Red Cross was already damaged to a certain extent.

Objectively speaking, not all the donations from Hong Kong were wasted. In April 2014, the Sichuan authorities announced that the donations from Hong Kong had HK$2 million left after the money had been spent on the buying of emergency equipment and facilities (*Apple Daily*, April 20, 2014: A14). The Sichuan provincial government also claimed that audit officials were responsible for inspecting how the donations were spent in the process of helping the rescue victims so that public accountability to outside donors, including the Hong Kong Jockey Club that donated HK$10 million to Sichuan, could be enhanced (*Ta Kung Pao*, April 20, 2014: A20). The Hong Kong Red Cross sent a delegation of 16 donors to visit Beichuan in May 2014 (*Sing Pao*, May 12, 2014: A19). It found that at least some of the HK$1.3 billion were worthwhile in assisting Beichuan's reconstruction of schools, houses and provision of rehabilitation services after the 2008 Wenchuan earthquake. However, some of the donations were misspent and wasted. In Guangyuan city, its deputy mayor Wu Lianqi was dismissed from both the Chinese Communist Party and public service after official investigation into his acceptance of bribes and "violation of the law" following the Wenchuan earthquake's relief measures (*Sing Tao Daily*, February 21, 2014: A24). In Mianyang city, a storage factory produced odious smell due to the rotten logistical supplies, including food, rice and water, which had been donated by outsiders. It was not known whether the rotten food and neglected water were donated from Hong Kong, but the party-secretary responsible for the storage supplies was ordered to terminate his work immediately. The lack of transparency in the mainland handling of the donations from outside remained an evidence of public maladministration, although outside donors could organize trips to inspect the situation of the affected earthquake regions.

PSYCHOLOGICAL COUNSELING

While the Tangshan earthquake in 1976 did not envisage post-earthquake reconstruction in the aspect of giving sufficient psychological counseling and services to the citizens, the 2008 Sichuan earthquake taught a lesson. Guangzhou's Youth Volunteers Association sent 100 psychological counselors who underwent special training to Sichuan in late May 2008 (*Sing Tao Daily*, May 28, 2008: B1). After an earthquake hit Japan's Kobe in January 1995 and killed 6,437 people, the Japanese government deployed psychological counselors to help the victims to recover quickly. Like the Kobe earthquake, the Sichuan earthquake prompted the PRC authorities at both central and provincial levels to pay special attention on how to recover the victims from their psychological shocks and pain.

According to Zhang Wei, the deputy director of the West China Hospital in Chengdu city, many earthquake victims and survivors suffered from post-traumatic stress disorder (PTSD), while others were too shy to seek help from any consultant. He said:

> The quake affected about 30 million people. With a prevalence rate of PTSD of up to 4 percent, there are more than 1 million patients out there. It's a large number . . . A typical symptom of PTSD is to often recall the disaster. For example, a woman who witnessed the death of her husband in the quake has continued to replay the scene in her mind, which is now seriously affecting her life and work . . . Many who suffer tend to regard the flashbacks they experience as natural memories of what happened, instead of what they really are, namely, psychological disorders . . . Students are vulnerable to the disorder because many of them witnessed the death of their peers. We hope that officials at all levels of government understand the importance of psychological consultation . . . Post-quake psychological reconstruction is never a short-term job. Without mental intervention, negative symptoms could last up to 20 years (*Beijing Review*, May 20, 2010).

He also revealed that many officials and police were vulnerable because they were under tremendous pressure of performing the dual roles of saving lives and aiding reconstruction. Apart from PTSD, many survivors exhibited symptoms of anxiety, hypersensitivity, sleeping disorders, and depression.

For a large number of earthquake survivors being traumatized, post-earthquake psychological counseling, assistance, and support were necessary.

CULTURAL HERITAGE PROTECTION

One of the most devastated areas of the Sichuan earthquake was the Qiang autonomous county in Beichuan, where one-tenth of members of the Qiang minority or 20,000 people died. The devastation was a huge blow to the cultural and historical development of the Qiang, whose cultural heritage villages in Mianyang and Beichuan were numerous (*Ming Pao*, June 22, 2008: A24). All the 800 pieces of cultural artefacts of Qiang were buried and destroyed in a museum showing the Qiang cultural heritage and historical uniqueness. In response, some cultural preservation experts and artists in Beijing went to Sichuan and discussed how the Qiang's cultural heritage and history would be resurrected in the affected regions. They also appealed to other artists and cultural workers to work together in Sichuan so as to preserve the unique historical and cultural history of the Qiang.

On the other hand, the police went to the Mianyang museum to collect all the national and local treasures and transport them to safe places. The Mianyang museum transferred 212 cases of cultural artefacts, including some 6,000 items of arts, paintings, porcelains, and historical documents (*World Journal*, May 30, 2008: B2). Due to the fact that a majority of cultural artifacts would be destroyed by water and flooding, the local police protected them and ensured that they would be transported and deposited in safe places.

CENTRAL PRESSURE ON THE ACCELERATION OF RECONSTRUCTION WORK

On March 5, 2009, Premier Wen delivered his work report and emphasized that reconstruction work after the Sichuan earthquake would have to be accelerated so that two rather than three years would realize the accomplishment of the recovery objective (Wang, 2013: 221). In response to the pressure from the central government in Beijing, the Sichuan provincial authorities delivered a report outlining all the achievements, such as the completion of one million buildings and housing units, the recovery of 748 urban facilities, the resumption of 738 public and cultural infrastructure projects, and the establishment of 1,433 cultural and market networks (Wang, 2013: 223). In Wenchuan county, a new Yingshou secondary school was constructed by using the donations from the special party dues of the CCP members in the entire nation (Wang, 2013: 225). The school's completion was a symbol of the success of the Party-state in the process of public mobilization in reconstruction.

The pairing of provincial and municipal partnership with the disaster-affected regions—a plan initiated by the central government in Beijing—did achieve impressive results. For instance, Shanghai helped Dujiangyan to complete various agricultural and science parks quickly, a demonstration that the former became the locomotive driving the speedy recovery of the latter (Wang, 2013: 229). Similarly, Zhejiang province helped the rapid reconstruction of Qingchuan and Guangyuan by developing ecological farmland and agricultural production. Liaoning also assisted An county to build up an industrial park and bring about enormous employment opportunities for the residents. The partnership between successful places and earthquake-wrecked regions proved to be a Chinese style of rapid recovery after the Wenchuan earthquake.

OFFICIAL PROPAGANDA LINE IN THE RECONSTRUCTION PHASE

The CCP identified the significance of using the Wenchuan earthquake's reconstruction process as a means to enhance the image of the ruling party. As such, since mid-June the Ministry of Propaganda began to hold a meeting among the official media organizations and editors. Presided over by the propaganda minister Liu Yunshan, the meeting emphasized the importance of highlighting the role of the CCP's leadership in the process of reconstruction, the planning on the part of the State Council, and the need to appeal to the masses for solidarity (*Sing Tao Daily*, June 16, 2008: B6). This directive from the central government in Beijing was a far cry from the relatively free-wheeling approach adopted in the first two weeks after the earthquake, which witnessed an unprecedented degree of openness, candid criticisms from the public, and transparency in the rescue operations as shown in the official mass media.

The official media also adopted the line of appealing to the parents of the dead children to avoid pursuing the question of responsibility in the "tofu" schools. A 45-year-old political activist and dissident, Huang Qi, opened a website in 1999 to advocate the importance of human rights and his website was changed to Tianwang Human Rights Affairs Center in 2006. Huang galvanized the grievances of the parents whose children died in collapsed school buildings in the Sichuan earthquake. But Huang and his workers were arrested in Chengdu in early June 2008 while his computer was confiscated by the police (*Sing Tao Daily*, June 16, 2008: B6). Obviously, Beijing and the provincial authorities in Sichuan attempted to tighten control over the mass media, hoping that the reconstruction phase would galvanize public support instead of allowing any political dissident to foster activities that would challenge the legitimacy of the CCP at the central and provincial levels.

In April and May 2012, some villagers and residents in Sichuan were critical of the local and provincial governments in Sichuan for failing to deal with reconstruction work properly and took to the streets. In April, some residents in Beichuan county complained that local government officials were corrupt and squeezed the outside donations to benefit themselves by being nontransparent in handling public accounts, by not resettling them to proper homes, and by terminating petitioners from going to Beijing to protest and disclose their misdeeds (*The Sun*, April 9, 2012: A26; *Oriental Daily News*, April 9, 2012: A22). The ways in which the Beichuan police stopped them from visiting Beijing to protest were an indication that local-level cadres and officials attempted to hide maladministration and corruption from the eyes and ears of the central government in Beijing. Meanwhile, several hundred villagers complained in public that their farmland was actually utilized by the village government in Wenchuan to build an industrial park rather than better homes for them (*The Sun*, May 13, 2012: A23). Those villagers who went to protest to the township government were arrested. Although the Wenchuan county government denied that there was any corruption on the part of its officials, the complaints from villagers tarnished the image that the Party-state was successful in dealing with reconstruction in all the affected regions after the Wenchuan earthquake.

Still, the CCP propaganda appeared to be quite successful in the reconstruction phase as the role of the Party was emphasized in many of the achievements at the local level in Sichuan province. For example, in the new Beichuan county, the CCP was described as having tremendous contribution to the region's recovery through the diligent work of its Party members and cadres in rescuing residents in the earthquake and providing leadership in reconstruction (Wang, 2013: 243–244). Parallel situation could be found in Yingshou, Qinghchuan, and Mianzhu. Wang Wenhong concluded in his work appreciative of the CCP efforts that although the Wenchuan earthquake brought about 69,227 deaths, 17,923 missing, and 400,000 injuries, costing a huge economic loss of 8,400 billion yuan, the reconstruction stage basically took only three years and its success represented the "power" of "the Chinese nation-state," the "Wenchuan spirit," and "the national spirit of the Great Wall" (Wang, 2013: 277–285). In February 2012, the provincial government in Sichuan claimed that 99 percent of all the reconstruction work was completed, involving an investment of 1.7 billion yuan (*Cheng Pao*, February 25, 2012: A15).

In the minds of most earthquake victims, according to a survey of 1,037 households conducted by two mainland Chinese scholars in six earthquake-affected regions in 2006, the local government that was perceived to have the strongest reconstruction capability was, in consecutive order, firstly Jiangyou, secondly Wenchuan, thirdly Shifang, fourthly Beichuan, fifthy Qingchuan, and finally Mianzhu (Chen and Meng, 2010: 41–42). Although public perceptions of the capability of local governments in reconstruction work varied, overall most citizens in affected regions were satisfied with the relatively rapid way in which the central, provincial, and local governments handled the relief and reconstruction work after May 12, 2008.

POLICY LEARNING FROM THE EARTHQUAKE, COOPERATION WITH TAIWAN AND BETTER DISASTER MANAGEMENT

In a sense, the PRC government and its provinces learned a lesson from the Wenchuan earthquake on how to handle another possible earthquake in the near future. In 2013, it was reported that 43,000 tremors occurred in Sichuan province alone (*The Sun*, January 4, 2014: A38). The policy learning dimensions of the Wenchuan earthquake could be seen in the process of rapid reconstruction and also the ways in which the PRC government later cooperated with the ROC counterpart on the detection of future earthquake. In February 2014, the PRC and ROC governments signed a formal agreement on the joint detection and information sharing of any future earthquake (*Ta Kung Pao*, February 28, 2014: A12). On September 21, 1999, Taiwan encountered an earthquake and its government since then has determined to make precautionary measures to tackle any other earthquake in the future. Hence, it senses the need to cooperate with seismological officials from mainland China in order to detect any earthquake and take prior preventive measures against natural disasters.

In reality, the PRC government has been determined to deal with natural disasters in a more effective manner since the 2008 Sichuan earthquake. It has tried to improve disaster response by building emergency supplies networks in various places. According to Li Baojun, the deputy director of the National Commission for Disaster Reduction, "We can distribute tents, food, drinking water, clothes and quilts to people affected in 12 hours or less after a natural disaster takes place" (*China Daily*, October 14: 2013: 1–2). He revealed the PRC established a four-tier disaster relief material storage network: 18 central-level disaster relief material bases in transport hubs nationwide and in all provincial capitals. Moreover, 93 percent of cities and 80 counties set up warehouses for disaster response in their regions. In October 2013, Typhoon Fitow hit the Zhejiang province, leading to the death of seven people and the need to allocate 1.24 million people to temporary shelters. The local government in Yuyao, which was a region worst hit by heavy rainfall in Zhejiang, was criticized by netizens for failing to provide immediate relief because many citizens were trapped in the flood and short of clean water for days. After the 2008 Sichuan earthquake and other natural disasters like typhoons and heavy snowfalls, the PRC government has been determined to consolidate its responses to natural disasters at all levels. In a sense, the concept of national security has been broadened to encompass the urgent need to tackle natural disasters in an effective and an efficient manner.

In May 2012, Sichuan's Chengdu city conducted an exercise among residents to deal with any possible earthquake. Simulated evacuations of schools children and citizens from their schools and homes were conducted; children were taught to hide under their tables in schools; and rescue workers were trained to save the lives of citizens who pretended to be trapped in a real earthquake (*ATV News*, May 30, 2012). Policy learning, in the minds of provincial and city officials in Sichuan, was necessary after the Wenchuan earthquake.

THE 2013 LUSHAN EARTHQUAKE IN CHINA AND COMPARISONS WITH THE 2008 WENCHUAN EARTHQUAKE

There were a number of earthquakes in the PRC after the 2008 Wenchuan disaster. On April 16, 2010, an earthquake of magnitude 7.1 took place in Yushu in the Tibet Autonomous Region and it killed 2,700 people. About 13,000 PLA soldiers and PAP police were mobilized quickly (*NIDS China Security Report* 2012, 2012: 23; *Beijing Review*, August 30, 2010). The PLA mobilization and command structure were similar to the situation in the Wenchuan earthquake. According to the Japanese Institute for Defence Studies, the PLA set up a similar command headquarters dealing with the Yushu earthquake and supervised the Lanzhou military region (*NIDS China Security Report 2012*, 2012: 23). It was clear that the PLA adopted a similar approach in dealing with the Yushu earthquake as with its previous experiences in tackling the Wenchuan tragedy. Although President Hu was visiting Brazil during the Yushu earthquake's occurrence, the State Council set up a parallel Chief Command Headquarters to cope with the natural disaster—a testimony of some degree of maturity on the part of the Party-state and military to tackle any earthquake after 2008 with ease and experiences.

On April 20, 2013, Sichuan's Yaan and Lushan counties were attacked by an earthquake of magnitude 7 that killed 200 people. President Xi Jinping quickly went to the disaster zone and encouraged residents to resist the earthquake and promised that the Party-state would reconstruct their homes immediately (*Hong Kong Commercial Daily*, May 22, 2013: A21). He also gave encouragement to the PLA, police, children, and peasants, demonstrating calmness and swift response. On May 2, Xi gave an order to the rescue operation saying that reconstruction would have to start immediately after the earthquake (*Wen Wei* Po, May 4, 2013: A8). In response to the Lushan earthquake, Beijing and the Sichuan provincial government provided a total of 1.51 billion yuan for rescue work, while every citizen in the quake-hit region received a subsidy of 10 yuan and 0.5 kg of grain per day, apart from the provision by civil affairs department of 87,000 tents, 233,000 quilts, and 3,393 tons of food to the affected regions (*China Daily*, April 28, 2013: 1). Although the Lushan earthquake destroyed some cultural heritage sites, the electricity supplies were resumed fully within a month (*Ming Pao*, April 27, 2013: A22). The ways in which the PRC government responded swiftly to the Lushan earthquake in 2013 was a testimony that it learned much from the 2008 earthquake in Sichuan and became very calm and effective in its responsive measures.

The 2013 Lushan earthquake did bring about dissident activities. According to reports from Hong Kong, the Lushan earthquake brought about a protest of some Chengdu citizens against the construction of a chemical factory for fear of gas leakage and environmental hazards (*Oriental Daily News*, April 28, 2013: A32). However, the Chendu government suppressed the protest and arrested its organizer. Obviously, the local government tended to be more politically conservative and cautious immediately after the earthquake for the sake of protecting its own image in the eyes of both the central government in Beijing and the foreign media. As with the 2008 Sichuan earthquake, some citizens affected by the 2013 Lushan earthquake criticized the seismological officials for failing to anticipate the disaster beforehand (*The Sun*, Aprl 27, 2013: 26). Again the seismological department defended by saying that it was difficult to predict the occurrence of an earthquake.

Huang Qi, a political dissident protesting against the "tofu" buildings after the 2008 Sichuan earthquake, was imprisoned for three years in November 2009 and released in June 2012, but he led 23 other dissidents to demonstrate against the "tofu" buildings again in the wake of the 2013 Lushan earthquake (*Hong Kong Daily News*, April 25, 2013: A10). He was then imprisoned again with three of his supporters on their way to Yaan county. Again, a similar pattern of political control on the part of local authorities could be seen in the aftermath of the 2013 Lushan earthquake, which stimulated a few political dissidents to protest against corruption and public maladministration.

Unlike the 2008 Wenchuan earthquake after which the PRC government received lots of donations and logistical support from other countries, the Foreign Ministry shortly after the 2013 Lushan earthquake expressed its gratitude to foreign countries for their concerns but added that it did not need the immediate assistance from their rescue teams due to the sufficient logistical supplies within China internally (*Hong Kong Economic Journal*, April 22, 2013: A20). The governments of the United States and Japan expressed their concerns and the PRC Ministry thanked them. The reactions of Beijing to the foreign concern showed that the PRC was more confident in handling a smaller scale earthquake in 2013, believing that its internal logistics and remedial measures would be sufficient to cope with the natural disaster.

The 2013 Lushan earthquake witnessed better PLA mobilization and response, perhaps a signal that the military also learned a lesson from the 2008 Wenchuan earthquake. As with Premier Wen in the 2008 Wenchuan earthquake, Premier Li Keqiang flew to the disaster regions five hours after the Lushan earthquake with his subordinates like Vice Premier Wang Yang and military generals, ordering that the PLA soldiers would have to refrain from sleeping so as to rescue the victims (*Hong Kong Daily News*, April 21, 2013: A1; *China Daily*, April 21, 2003: 1). Li remarked:

> Currently, the most urgent issue is to save people. We cannot waste a single second in the first 72 hours after the quake, which is the peak time for saving lives. First we have to rescue the trapped from the debris and then check the quake-hit buildings, one by one. We have experienced disasters. We know how to cope with it, and we have an emergency mechanism. We have faith in our ability to overcome the difficulties, minimizing the losses under the leadership of the Party and the government (*China Daily*, April 21, 2013: 4).

Immediately after the Lushan earthquake, the PLA, PAP, police, and relief personnel were mobilised and deployed to the scene with plentiful relief supplies (*South China Morning Post*, April 21, 2013: 6–7). The National Earthquake Administration, the Ministry of Civil Affairs, public security, and government-sponsored charity organizations came to the rescue and relief operations quickly. About 7,500 PLA soldiers and PAP police were immediately sent to the disaster zone and some 10,000 others were placed on standby situation. The commander of the air force, General Ma Xiaotian, arrived in Sichuan's Qionglai county to set up a headquarters for the rescue and relief operations. He ordered two reconnaissance aircrafts from Beijing to take aerial photographs of the quake-hit regions, whereas five civilian drones were sent to inspect the seriousness of the earthquake's impacts. The reconnaissance appeared to be much faster and more effective than the 2008 Wenchuan earthquake. Forty-five civilian flights were chartered to transport logistical supplies to the disaster zones, but a military vehicle carrying 17 soldiers had an accident that killed two of them. The PRC Red Cross utilized its warehouse in Chengdu to deliver logistical supplies, including 1,700 tents, to the affected areas. Volunteer groups such as the Chongqing Huayan Fund for Culture and Education sent volunteers to the scene quickly. Overall, the responses from the military, central and local governments, and the private sector were swift. Tang Yu, a worker of the Chengdu branch of the Adventist Development and Relief Agency in China remarked:

> It's best for the government to lead relief operations because they have the resources and the power of mobilisation if they can deliver these in an efficient

> way . . . NGOs like us can at best facilitate the government's efforts because we have limited manpower and resources (*South China Morning Post*, April 21, 2013: 6–7).

On February 19, 2013, an earthquake of magnitude 4.9 took place in Yunnan province, but the official authorities said that they had predicted its occurrence beforehand (*Oriental Daily News*, September 9, 2012: A28). Still, the earthquake brought about a death toll of 80 citizens. In July 2013, Gansu province's Min county was affected by an earthquake of magnitude 6.6 (*Oriental Daily News*, July 29: 2013: A26). The series of earthquakes in the PRC after the 2008 Wenchuan earthquake did not lead to serious death tolls and casualties, partly because of their small magnitude and mainly because the PRC government at both central and provincial levels have learned lessons on how to cope with natural disasters in a more effective manner since the Wenchuan earthquake on May 12, 2008.

On May 12, 2014, which was the sixth anniversary of the Wenchuan earthquake, a ceremony was held in Yingshou county where 300 residents went to pay tribute to their lost relatives in a public cemetery (*Sing Pao*, May 13, 2014: A20). On the same day, Wenchuan's fire services and rescue teams conducted a simulation exercise in preparation for any reoccurrence of an earthquake. The psychological and practical memories of the Wenchuan earthquake are deep in the psyche of residents and public servants in Sichuan province.

However, a dark side of the sixth anniversary of the Wenchuan earthquake was that political control of dissidents persisted in the PRC. Tan Zuoren, a dissident who was critical of the substandard schools and buildings in Wenchuan and who openly wrote articles in memory of the tragedy on June 4, 1989 on the Tiananmen Square, was released from prison by the Sichuan authorities in March 2014, but he was warned of the dangers of contacting foreign media (*Oriental Daily News*, May 12, 2014: A26). Tan's name could not be searched in the mainland's blogs and websites, reflecting the state's censorship of the mass media. Rather, in Sichuan's media, terms like "new phenomena," "*taohua yuan*" (flower lake) and "water and fire reborn" were used to describe the regions affected by the earthquake (*Oriental Daily News*, May 12, 2014: A26). Clearly, the state propaganda was intermingled with selective political control to portray an image of a renewed legitimacy enjoyed by the provincial authorities.

CONCLUSION

The Party-state utilized the reconstruction phase after the Wenchuan earthquake to mobilize the public, the provinces, and cities to assist the disaster-plagued regions. The result was prominent: central directive and mobilization of the public, provinces, and localities did contribute to the relatively rapid recovery of the affected regions in Sichuan and other places. As a result, the CCP and the PRC government were successful in turning the earthquake into a golden opportunity of winning the hearts and minds of the Chinese citizens. The legitimacy of the Party-state was therefore enhanced through the entire reconstruction process. Finally, the concept of national security has been broadened to encompass the need for the Party-state to respond to natural disasters, such as earthquake and typhoons, in a far more effective and efficient manner than ever before. This broadened concept of national security has become very prominent in the wake of the post-earthquake reconstruction and precautionary work.

Chapter Five

Crisis Management in Comparative Perspective

This chapter adopts a comparative perspective to analyze the responses of the PRC to crises, such as Tangshan and Wenchuan earthquakes and other natural disasters like the Severe Acute Respiratory Syndrome (SARS) in 2003, in comparison with the ways in which other Asian states cope with crises. First, it will review how the PRC government tackled the Tangshan earthquake in 1976 and cast light on the comparisons and contrasts with how it dealt with the 2008 Wenchuan earthquake. Second, it will examine the eruption of SARS in 2003 and analyze any possible progress in China's handling of crises in light of the responses of the Party-state and military to the 2008 Wenchuan earthquake, particularly if crises embrace both natural disasters like earthquake and contagious diseases as with SARS. Third, this chapter will compare and contrast the PRC responses to the 2008 Wenchuan earthquake with the 2010 Yushu earthquake in Qinghai province and the 2013 Yaan earthquake in Sichuan. Fourth, it will explore how Myanmar tackled the May 2008 cyclone in comparison and contrast with the PRC responses to the Wenchuan earthquake that occurred in the same month. Fifth, it will examine the similarities and differences between the Chinese handling of the Wenchuan earthquake with the Japanese tsunami in 2011. Sixth, some contrasts with the Chile earthquake in 2010 and 2014 will also be made. Finally, it will draw conclusions on the Chinese crisis management from a comparative vantage point of view.

THE 1976 TANGSHAN EARTHQUAKE: CONTRASTS WITH THE 2008 WENCHUAN EARTHQUAKE

There were many earthquakes in the world (see table 5.1). In China, on July 28, 1976, the Tangshan earthquake of magnitude 7.8 took place and it killed 242,769 people as well as injuring 164,851 residents (Qian, 2010: 12). According to an eyewitness account by Qian Gang, who experienced the tragedy at that time, most buildings in Tangshan were destroyed; railway lines were totally distorted; landscape was altered; and the whole region was almost flattened. A week before the earthquake, residents in Tangshan had already found that an unprecedented number of fish surfaced on the ponds and rivers, allowing fishermen to catch them easily (Qian, 2010: 7). Rats and wolves even ran away from their caves, appearing to signal a forthcoming ominous event. Dogs, chickens, and horses also howled, scaring the Tangshan residents (Qian, 2010: 14). On the morning of July 28 at 3:42 AM, the earthquake attacked the city, destroying one of the largest industrial cities in Northern China. Tangshan was famous for its coal production, porcelain products, mining, textile industry, mechanical engineering sector, and automobile industry. Once the earthquake occurred, the city officials requested that Beijing should send the military, health teams, and mining rescue workers to the disaster scene. At that time, the Shenyang military region's General Li Desang was responsible for dispatching the PLA to the affected regions, while the Beijing military's deputy commander Li Min arrived at Tangshan on the morning of July 28 at 10 AM. They were followed by the dispatch of the PLA and rescue teams from the southwest and northeast of Tangshan (Qian, 2010: 48–49). Fortunately, the Tangshan earthquake did not severely bring about other natural disasters, such as gas leakage, because repair work on a chemical factory in Kaiping stopped further leakage of chemical and poisonous gas. Nevertheless, many mine workers were trapped and buried.

Fourteen hours after the earthquake, the first PLA division from Shenyang arrived at the city (Qian, 2010: 61). In total, the central government and Shenyang military region sent 100,000 PLA soldiers to participate in the rescue operation. However, death tolls in Tangshan were serious. Although some citizens volunteered to help each other, many rescue workers and residents died of accidents. The 225 Military Hospital was full of hundreds of thousands of injured citizens; almost one out of five Tangshan residents was seriously injured, amounting to 100,000 people (Qian, 2010: 73). On 30 July, the State Council decided to airlift the injured citizens to 11 other provinces and cities, but the number of flights per day from 30 July to 5 August was between 12 and 58, transporting a range of 500 to 4,500 citizens out of the city every day (Qian, 2010: 75). PLA planes had to remove all the seats in order to accommodate the injured citizens, but still the number of planes and helicopters was insufficient. Sometimes it took two hours for injured resi-

Table 5.1 Major Earthquakes After the Second World War

Origin	*Location*	*Magnitude*
1946/04/01	Unimak Island, Alaska	8.1
1946/08/04	Samana, Dominican Republic	8.0
1946/08/08	Samana, Dominican Republic	7.6
1946/12/20	Nankaido, Japan	8.1
1949/07/10	Khait, Tajikistan	7.5
1949/08/22	Queen Charlotte Islands, British Columbia, Canada	8.1
1950/08/15	Assam - Tibet	8.6
1952/03/04	Hokkaido, Japan region	8.1
1952/11/04	Kamchatka	9.0
1954/03/29	Spain	7.9
1957/03/09	Andreanof Islands, Alaska	8.6
1957/06/27	Stanovoy Mountains, Russia (USSR)	7.6
1957/07/28	Guerrero, Mexico	7.9
1957/12/04	Gobi-Altay, Mongolia	8.1
1958/07/10	Lituya Bay, Alaska	7.7
1958/11/06	Kuril Islands	8.3
1959/04/26	Taiwan region	7.5
1960/01/13	Arequipa, Peru	7.5
1960/05/21	Arauco Peninsula, Chile	7.9
1960/05/22	Chile	9.5
1963/10/13	Kuril Islands	8.5
1964/03/28	Prince William Sound, Alaska	9.2
1964/06/16	Niigata, Japan	7.5
1965/01/24	Sanana, Indonesia (Ceram Sea)	7.6
1965/02/04	Rat Islands, Alaska	8.7
1965/03/14	Hindu Kush, Afghanistan	7.8
1966/10/17	Near the Coast of Peru	8.1
1968/05/16	Off the East Coast of Honshu, Japan	8.2
1969/02/28	Portugal-Morocco area	7.8
1970/01/04	Yunnan Province, China	7.5
1970/05/31	Chimbote, Peru	7.9
1970/07/31	Colombia	8.0
1971/07/09	Valparaiso region, Chile	7.5

1972/01/25	Taiwan region	7.5
1972/07/30	Sitka, Alaska	7.6
1974/10/03	Near the Coast of Central Peru	8.1
1974/10/08	Leeward Islands	7.5
1975/02/02	Near Islands, Alaska	7.6
1976/07/27	Tangshan, China	7.5
1976/08/16	Mindanao, Philippines	7.9
1978/09/16	Iran	7.8
1979/02/28	Mt. St. Elias, Alaska	7.5
1980/10/10	El Asnam (formerly Orleansville), Algeria	7.7
1985/03/03	Offshore Valparaiso, Chile	7.8
1985/09/19	Michoacan, Mexico	8.0
1987/11/30	Gulf of Alaska	7.8
1988/03/06	Gulf of Alaska	7.7
1990/07/16	Luzon, Philippine Islands	7.7
1991/04/22	Costa Rica	7.6
1992/12/12	Flores Region, Indonesia	7.8
1993/08/08	South of the Mariana Islands	7.8
1994/06/09	Bolivia	8.2
1996/06/10	Andreanof Islands, Alaska	7.9
1997/10/14	South of Fiji Islands	7.8
1997/12/05	Near East Coast of Kamchatka	7.8
1998/01/04	Loyalty Islands Region	7.5
1998/03/25	Balleny Islands Region	8.1
1998/05/03	Southeast of Taiwan	7.5
1999/08/17	Izmit, Turkey	7.6
1999/09/20	Taiwan	7.6
1999/09/30	Oaxaca, Mexico	7.5
2000/06 /04	Southern Sumatera, Indonesia	7.9
2000/06/18	South Indian Ocean	7.9
2000/11/16	New Ireland Region, Papua New Guinea	8.0
2000/11/16	New Ireland Region, Papua New Guinea	7.8
2000/11/17	New Britain region, Papua New Guinea	7.6
2001/01/01	Mindanao, Philippines	7.5
2001/01/13	El Salvador	7.7
2001/01/26	Gujarat, India	7.6

2001/06/23	Near the Coast of Peru	8.4
2001/07/07	Near the Coast of Peru	7.6
2002/03/05	Mindanao, Philippines	7.5
2002/08/19	Fiji Islands	7.7
2002/09/08	New Guinea, Papua New Guinea	7.6
2002/10/10	Irian Jaya, Indonesia	7.6
2002/11/03	Denali Fault, Alaska	7.9
2003/01/22	Offshore Colima, Mexico	7.6
2003/07/15	Carlsberg Ridge	7.6
2003/08/04	Scotia Sea	7.6
2003/09/25	Hokkaido, Japan Region	8.3
2003/11/17	Rat Islands, Aleutian Islands, Alaska	7.8
2004/11/11	Kepulauan Alor, Indonesia	7.5
2004/12/23	North of Macquarie Island	8.1
2004/12/26	Sumatra-Andaman Islands	9.1
2005/03/28	Northern Sumatra, Indonesia	8.6
2005/06/13	Tarapaca, Chile	7.8
2005/09/09	New Ireland Region, Papua New Guinea	7.6
2005/09/26	Northern Peru	7.5
2005/10/08	Pakistan	7.6
2006/01/27	Banda Sea	7.6
2006/04/20	Koryakia, Russia	7.6
2006/05/03	Tonga	8.0
2006/07/17	South of Java, Indonesia	7.7
2006/11/15	Kuril Islands	8.3
2007/01/13	East of the Kuril Islands	8.1
2007/01/21	Molucca Sea	7.5
2007/08/08	Java, Indonesia	7.5
2007/08/15	Near the Coast of Central Peru	8.0
2007/09/12	Southern Sumatra, Indonesia	8.5
2007/09/12	Kepulauan Mentawai region, Indonesia	7.9
2007/09/28	Mariana Islands region	7.5
2007/11/14	Antofagasta, Chile	7.7
2008/07/05	Sea of Okhotsk	7.7
2009/01/03	Near the North Coast of Papua, Indonesia	7.7
2009/03/19	Tonga region	7.6

2009/07/15	Off West Coast of the South Island, New Zealand	7.8
2009/08/10	Andaman Islands, India region	7.5
2009/09/29	Samoa Islands region	8.1
2009/09/30	Southern Sumatra, Indonesia	7.5
2009/10/07	Vanuatu	7.7
2009/10/07	Santa Cruz Islands	7.8
2010/02/27	Offshore Bio-Bio, Chile	8.8
2010/07/23	Moro Gulf, Mindanao, Philippines	7.6
2011/03/11	Near the East Coast of Honshu, Japan	9.0

Source: http://earthquake.usgs.gov/earthquakes/world/historical.php, accessed on October 13, 2011.

dents to get on a PLA plane because the ladder from the ground to the plane's door was relatively high and thus hindering the transportation process (Qian, 2010: 76). On July 13, the Beijing military region concluded that 100,000 personnel from the air, land, and naval forces in Beijing and Shenyang participated in the rescue operations and that 12,245 residents were rescued from July 28 to August 10 (Qian, 2010: 80). The rescue tools were relatively rudimentary, including shovels, cranes, and electrical cutters.

Some foreigners who worked in Tangshan died, including three of the nine Japanese technicians (Qian, 2010: 119). On July 28, the Foreign Ministry quickly sent a plane to evacuate the foreign residents from Tangshan. Rescue workers in the city included not only the military, police, and health personnel but also prisoners who also participated in the rescue operations. Although a mental hospital was flattened, many citizens suffered from the trauma, and from psychological shocks and depression, with some accompanied by relatives to visit the collapsed mental hospital for assistance (Qian, 2010: 137).

Crime was rampant in the city after the earthquake. Some robbers snatched any goods they could find in department stores, including seafood, batteries, and radios. Civilian police reported that in the first week after the earthquake 1,800 criminal elements were arrested for stealing logistical supplies and precious belongings like watches and cash (Qiao, 2010: 161). Immediately after the earthquake, police were sent to guard the warehouses and banks so as to deter any robbery. On his way to visit the rescue team and PLA soldiers, the Beijing military region's deputy director Chi Haotian saw a group of citizens beating up a robber and the Hebei province's party-secretary Liu Zihou remarked that the local government allowed such vigilante groups to maintain law and order. In view of rampant criminal activities, the Tangshan city's Rescue Command Headquarters issued Directive Number 1, which was posted on the walls and streets, saying that (1) citizens were not allowed to engage in theft, (2) all logistical supplied stolen had to be returned, and (3) the civilian police who maintained law and order were not allowed to open fire arbitrarily (Qian, 2010: 167).

Infectious diseases broke out because viruses emerged from human feces, flies, pollution, dead animals, and corpses amid high temperatures. Beijing set up a leading small group on health maintenance in Tangshan and dispatched 21 health teams with 1,300 personnel staff members spraying anti-decay chemicals across the earthquake-hit city (Qian, 2010: 171). Masks had to be required for all health workers handling dead bodies and funeral procedures were forced to be simple by burning the corpses or using mass burial.

The PRC government at that time denied any assistance offered by foreign countries, such as the United States, the United Kingdom, and Japan, for the sake of maintaining its own self-sufficiency and national dignity (Qian, 2010: 187–189). Some CCP cadres asserted that the Tangshan earthquake provided an opportunity for the "education of communism," meaning that the PRC regime led by Chairman Mao Zedong could lead the masses to prevail over the natural disaster (Qian, 2010: 192–193). By September 1976, the third and final batches of orphans were transported out of Tangshan city with the involvement of the transmission belts of the CCP, such as the Federation of Women, the Federation of Trade Unions, the Communist Youth League, and the Ministry of Civil Affairs.

Long before the Tangshan earthquake, Premier Zhou Enlai had appealed to seismological officials to detect any possible earthquake pattern as early as March 8, 1966 when an earthquake of magnitude 6.8 broke out in Xintai. However, officials responsible for earthquake detection appeared to find it difficult to predict the Tangshan earthquake although they tried their best to do so scientifically. Qian Gang documented that seismological officials did predict the imminent occurrence of an earthquake in the PRC two weeks before the Tangshan tragedy, but the Beijing leaders were relatively slow in taking precautionary measures and issuing warnings to the anticipated locations of such earthquake (Qian, 2010: 248–253).

There were some similarities and considerable differences between the governmental handling of the 2008 Wenchuan earthquake and that of the 1976 Tangshan one. In terms of the most important similarity, the seismological officials appeared to be able to predict the 1976 Tangshan earthquake, but the predictions came relatively so late that Beijing and Hebei province were unable to respond swiftly. Both earthquakes necessitated immediate mobilization of the military by the central government in Beijing. In contrast, the seismological officials were unprepared and unable to anticipate the Sichuan earthquake in 2008. The differences, however, were startlingly prominent. The PLA responses were apparently more mobile and much swifter in the 2008 Sichuan earthquake than the situation in 1976, although Premier

Wen was still dissatisfied with the reaction and performance of the military. The PLA equipments and rescue tools appeared to be far less advanced and more rudimentary after the 1976 Tangshan earthquake.

Furthermore, criminal activities appeared to be far more serious in the 1976 Tangshan earthquake than the 2008 circumstances. Nor did the NGOs emerge in 1976, unlike 2008 when NGOs emerged after the Wenchuan earthquake. The mass media were relatively underdeveloped in the PRC in 1976, unlike 2008 when the official Chinese media were mobilized swiftly as an effective tool to expose the tragedy to the audience in all mainland provinces and cities. Foreign and Hong Kong media could also cover the 2008 post-earthquake relief work easily, but they had great difficulties in entering the PRC to cover the impacts of the 1976 Tangshan earthquake. Full provincial mobilization in assisting Wenchuan was evident in 2008, but the central government appeared to mobilize some provinces and cities in the 1976 Tangshan post-earthquake relief and rescue work. Crime was rampant in 1976, but criminal activities were by and large kept under control after the 2008 Wenchuan earthquake. Foreign aid was rejected by the self-sufficient PRC regime in 1976, whereas the PRC government under reform era embraced and welcomed foreign assistance after the 2008 earthquake. Overall, in terms of the governmental capability of coping with rescue, relief, and reconstruction work, the PRC regime after the 2008 disaster was far more effective than that in 1976. Indeed, with the imminent holding of Olympics in the PRC, the Chinese Party-state performed in a way very different from its counterpart in the Maoist era. The new Chinese Party-state in 2008 had to project an international image of a competent and an effective administration with less than three months away from the Olympics. Hence, the world was watching how the Party-state in the PRC handled the legacies and impacts of the 2008 earthquake, unlike 1976 when the PRC was relatively closed in its domestic and foreign policies. It can be argued that the impacts of the 1976 Tangshan earthquake on the legitimacy of the CCP was minimal, for the social forces like NGOs at that time were almost nonexistent, not to mention any challenge to the Chinese state. Furthermore, the top-down nature of emergency relief operations in 1976 meant that citizens had no choice but to rely on the Party-state for rescue, relief, and reconstruction. In 2008, however, the rapid responses from Beijing and the mobilization of provinces and cities, together with the emergence of NGOs supportive of the governmental efforts at conducting the rescue, relief, and reconstruction work, led to a high degree of legitimacy at both the central and provincial levels. Although some citizens did protest against the "tofu" buildings in Sichuan after the earthquake, they represented a minority of citizens and did not really pose a challenge to the legitimacy of the Party-state in China.

THE 2003 SEVERE ACUTE RESPIRATORY SYNDROME (SARS)

In December 2002, it was reported in Hong Kong that there was a mysterious disease named atypical pneumonia and that it also occurred in Southern China where citizens in the city of Guangzhou scrambled for vinegar with a view of killing germs. At that time, it was unknown that the disease later turned out to be the Severe Acute Respiratory Syndrome (SARS). If infectious disease is defined broadly as viruses that can be transmitted either from animals to human beings or from humans to humans, it embraces HIV/AIDS, venereal disease, bird flu, West Nile virus and SARS.

Table 5.2 shows that the PRC had the largest number of people infected with SARS. Although their case fatality ratio was claimed to be 7 percent, 349 people died of the disease. Due to the relative absence of information openness on the part of the PRC government from January to March 2004, the official fatality figure might not be totally accurate. Other Asian cities and states that envisaged a relatively high ratio of case fatality included Hong Kong, Singapore, Thailand, Taiwan, Malaysia, the Philippines, and Vietnam. The serious situation reflected the virulent nature of SARS.

During the early phase of the outbreak of SARS from December 2002 to March 2003, Beijing encountered a cover-up problem of local governments in Guangdong province, which appeared to hide any negative news prior to the annual National People's Congress meeting in the capital. In April 2003, when President Hu Jintao dismissed the minister of health and the Beijing mayor for mishandling SARS, the provincial and local governments became aware of the importance of responding to the infectious disease quickly. In addition, Beijing fully mobilized the official mass media to address the SARS problem. Partly because of the need to protect its international image as many countries in the world wondered what happened with the public health circumstances in the PRC, and partly due to the visit of the World Health Organization (WHO) officials to Beijing to look into the situation, the PRC government had no choice but to release more information to the outside world while utilizing information openness to exert pressure on all the provincial and local administrations to tackle SARS in a more effective way (Saich 2006; Lo, 2007). In fact, when the WHO delegation visited Beijing's 301 Military Hospital, it was reported in Hong Kong that some SARS patients were already transferred to other places. The WHO eventually issued travel advisories to persons who visited both Hong Kong and the PRC and recommended them to consider the possibility of delaying their trips in April and May 2003. The outbreak of SARS in Hong Kong was attributable to an index patient, who was a doctor visiting the territory from mainland China and who infected many other people in Hong Kong during his stay in a hotel and later a hospital.

Table 5.2 SARS and Its Impacts on States and Cities, November 1, 2002 to July 31, 2003

Areas	*Cumulative number of cases*			*Number of deaths*	*Case fatality ratio (%)*	*Date onset first probable case*	*Date onset last probable case*
	Female	*Male*	*Total*				
Australia	4	2	6	0	0	26/2/2003	1/4/2003
Canada	151	100	251	43	17	23/2/2003	12/6/2003
China	2674	2607	5327	349	7	16/11/2002	3/6/2003
Hong Kong	977	778	1755	299	17	15/2/2003	31/5/2003
Macao	0	1	1	0	0	5/5/2003	5/5/2003
Taiwan	218	128	346	37	11	25/2/2003	15/6/2003
France	1	6	7	1	14	21/3/2003	3/5/2003
Germany	4	5	9	0	0	9/3/2003	6/5/2003
India	0	3	3	0	0	25/4/2003	6/5/2003
Indonesia	0	2	2	0	0	6/4/2003	17/4/2003
Italy	1	3	4	0	0	12/3/2003	20/4/2003
Kuwait	1	0	1	0	0	9/4/2003	9/4/2003
Malaysia	1	4	5	2	40	14/3/2003	22/4/2003
Mongolia	8	1	9	0	0	31/3/2003	6/5/2003
New Zealand	1	0	1	0	0	20/4/2003	20/4/2003
Philippines	8	6	14	2	14	25/2/2003	5/5/2003
Republic of Ireland	0	1	1	0	0	27/2/2003	27/2/2003
Republic of Korea	0	3	3	0	0	25/4/2003	10/5/2003
Romania	0	1	1	0	0	19/3/2003	19/3/2003
Russian Federation	0	1	1	0	0	5/5/2003	5/5/2003
Singapore	161	77	238	33	14	25/2/2003	5/5/2003
South Africa	0	1	1	1	100	3/4/2003	3/4/2003
Spain	0	1	1	0	0	26/3/2003	26/3/2003
Sweden	3	2	5	0	0	28/3/2003	23/4/2003
Switzerland	0	1	1	0	0	9/3/2003	9/3/2003
Thailand	5	4	9	2	22	11/3/2003	27/5/2003

United Kingdom	2	2	4	0	0	1/3/2003	1/4/2003
United States	13	14	27	0	0	24/2/2003	13/7/2003
Vietnam	39	24	63	5	8	23/2/2003	14/4/2003

Source: Modified table from http://www.who.int/csr/sars/country/table2004_04_21/en/index.html

Due to the smallness of Hong Kong, any infectious disease would harm the population's health quickly, thus prompting the territory to enhance communication and cooperation with both the PRC and WHO to stem the spread of SARS. As such, Hong Kong's officials segregated the potential SARS patients in dormitories located in remote areas while installing infrared temperature testing instruments at the international airports and ports so as to identify potential virus carriers and then isolate them geographically.

In the PRC, however, the vastness of the country meant that hospital and customs officials did not adopt stringent measures as with the Hong Kong case. While the mass media in Hong Kong aggressively took the government to task and made it more accountable to the public for the daily casualties and death tolls resulting from SARS, the mainland media lacked the checks and balances power of the Hong Kong counterpart.

The responses of the PRC government to SARS could be divided into two stages, including the earlier phase during which local protectionism and unwillingness to report any negative news brought about the rapid spread of SARS, and the later phase in which the central government in Beijing was determined to mobilize all the provinces, cities, and official media to curb the advance of SARS.

Comparatively speaking, the PRC state's responses to SARS were different from its reactions to the 2008 Wenchuan earthquake in several aspects. First and foremost, the 2008 Sichuan earthquake caused the prominent deaths of Chinese citizens and pushed the governments at all levels to respond without any delay; nevertheless, the emergence of the mysterious disease, the ignorance of its identity and nature, and the tendency of local authorities to refrain from reporting negative news prior to the annual meeting of the National People's Congress were the three main factors contributing to the relatively sluggish response of the Party-state to SARS from December 2002 to March 2003. Second, the nature of infectious disease is a far cry from sudden natural disasters like earthquake; SARS required some time for the authorities to diagnose the new virus but earthquake really demanded that the PRC government should respond promptly and effectively. Third, while the imminent Olympics Games in Beijing necessitated the central government to tackle the impacts of the Wenchuan earthquake immediately and effectively, there was no time incentive in the provincial and local governments' handling of the mysterious disease from December 2002 to March 2003. Fourth, while the mass media were encouraged by the central and provincial governments to tackle the Wenchuan earthquake's rescue and relief operations collectively, they were under the tight control of Beijing until the green light was issued by President Hu in March 2003. Fifth, the concerns of foreign countries and the WHO exerted pressures on Beijing to handle the spread of SARS to other parts of the world promptly, but similar concerns actually facilitated the international assistance of Beijing and Sichuan province to deal with the rescue and relief operations more successfully. Sixth, the two crises in 2003 and 2008 demonstrated that the PRC government was both calm and resilient in handling them, demonstrating the maturity of the Party-state in crisis management, although there were differences in how it handled the two events as mentioned before.

THE YUSHU EARTHQUAKE IN 2010 AND THE YAAN EARTHQUAKE IN 2013

On April 14, 2010, an earthquake of magnitude 7.1 hit the Yushu county in Qinghai province, killing 2,698 people, injuring at least 10,000 citizens, and prompting Vice Premier Hui Liangyu to visit the affected region immediately (*Xinhua*, April 15, 2010; *Xinhua*, June 2, 2010). Hui commanded the State Council's General Headquarters for Earthquake Relief and was assisted by the PLA and PAP to conduct the rescue operation at once (Mulvenon, June 2010: 3). On April 15, Premier Wen went to the county and inspected the situation, emphasizing that the central government shared the pain of the victims and would continue to try its utmost to provide relief to those who lost their loves ones (*CCTV*, April 17, 2010). On April 17, once President Hu Jintao returned to China from Latin America, he held a Politburo Standing Committee meeting to hear the reports on the earthquake. On April 18, he visited the affected regions amid rumors that the high-profile approach that had been adopted by Premier Wen in the 2008 Wenchuan earthquake might anger him (Mulvenon, June 2010: 3).

A total of 15,000 rescuers, including 11,000 soldiers from the PLA, PAP, firefighters, and special police, were deployed from the Lanzhou military region at once after the earthquake. The Lanzhou military command received the immediate order from the Central Military Commission and PLA's General Staff Headquarters, which received a report of the earthquake 12 minutes after it broke out. The coordinating work of the PLA was the responsibility of the Office of the PLA Leading Group for Handling Emergencies under the Central Military Commission (Mulvenon, June 2010: 4). In Yushu, there was no railway and only an airport and 90 percent of the residents were Tibetans.

Major General Wang Zhenguo announced in a press conference that many rescuers suffered from altitude sickness and two soldiers who developed pulmonary edema had to receive medical treatment (*China Daily*, April 20, 2010). Due to the sensitivity of the affected region where many Tibetans resided, President Hu on April 18 arrived at the disaster areas and promised the victims that the government would rebuild homes for them (*Xinhua*, April 18, 2010). Tents, water, and food were immediately provided for the victims by the Ministry of Civil Affairs, whereas tons of quilts and coats were transported to the affected regions. Although the rugged terrains and the high altitudes accompanied by the thin air in Yushu, which was about 4,000 meters above the sea level, made the rescue operation relatively difficult, the PRC government was swift in mobilizing the military and departments concerned to conduct the rescue operation. While some citizens volunteered to rush to the scene and join the rescue operation, the government authorities persuaded them not to do so because of the freezing weather conditions and the high altitude environment (*CCTV*, April 19, 2010). From mid-April to December 2010, some charity organizations went to Yushu to provide various relief support for the earthquake victims, including the Jinde Charities, the Xian Catholic Social Service Center, the Xining Diocese, the Xining Parish, the Lanzhou Catholic Social Service Center, the Guangxi Social Service Center, the Ci Xingxiyuan Orphanage, the Kungpin Vocational Training Center, and the Yushu Buddhist College (Jinde Charities, 2010: 1–16). On June 2 2010, Xi Jinping, the PRC vice-president, called on the CCP to take the lead in the reconstruction work of Yushu (*Xinhua*, June 2, 2010). He went to visit the Yushu Hongqi Elementary School and praised the work of the PLA and the PAP in conducting the rescue operation successfully.

On August 1, 2012, when the PLA celebrated its founding every year during the 1927 Nanchang uprising, the official mouthpiece, *People's Daily*, emphasized the success of military in "serving" both the party and the state and also its achievement in dealing with the rescue operations immediately after the Wenchuan and Yushu earthquakes (*People's Daily*, August 1, 2012; Chase, 2012). At the same time, *Xinhua* commented that the PLA would have a crucial role to play in defending national security (*Xinhua,* July 31, 2012), which has since the 2008 Wenchuan earthquake been broadly defined to encompass not only traditional security but also other areas such as public health, environment, and natural disasters. Clearly, the CCP legitimized the performance of the PLA by referring to its successful rescue operations in the aftermath of the Wenchuan and Yushu earthquakes, unintentionally or intentionally reasserting the military authority after the June 1989 Tiananmen tragedy during which the PLA had to be mobilized by the top CCP leaders to quell the student democrats calling for democracy on the Tiananmen Square.

Comparatively speaking, the role of the premier was slightly different in the Wenchuan and Yushu earthquakes. Perhaps because the State Council accumulated much experience in dealing with the rescue operation during the 2008 Wenchuan earthquake's rescue operation, Premier Wen first delegated the responsibility of leadership and coordination to Vice Premier Hui Liangyu. Although Wen did visit the affected regions on the second day of the earthquake, the situation appeared to be less tense than the 2008 Wenchuan disaster. Judging from the ways in which the government and nongovernmental organizations reacted quickly shortly after the Yushu earthquake, the PRC has since 2008 built up its capability of coping with the aftermath of any earthquake.

On April 20, 2013, Sichuan's Yaan and Lushan cities witnessed an earthquake of magnitude 7 that killed about 200 residents and injured 11,000 others *(Wen Wei Po*, April 21, 2014: A14; *The Standard*, April 22, 2013: 1). Many citizens were trapped in rubble. Learning from the 2008 Wenchuan earthquake after which volunteers and groups going to rescue victims unintentionally blocked many roads that even affected the rescue operation conducted by the PLA, the State Council in the aftermath of the April 2013 Yaan earthquake did not allow unauthorized rescue efforts and individuals from going into the affected areas for fear of hampering traffic (*The Standard*, April 22, 2013: 1). Within eight hours after the earthquake, Premier Li Keqiang and Vice Premier Wang Yang, together with high-ranking commanders of the PLA, arrived at the earthquake scene and 7,500 military soldiers were mobilized to conduct the rescue operation (*Sing Pao*, April 21, 2013: A2). Li called for the rescue operation to be stepped up before his return to Beijing (*The Standard*, April 22, 2014: 1). He said: "The most important task is to familiarize ourselves with the disaster conditions for coordinated rescue operations" (*The Standard*, April 22, 2013: 1). His reaction to the earthquake was much milder than his predecessor Wen Jiabao, who criticized the PLA for being sluggish in the rescue operation after the 2008 Wenchuan earthquake and who angered some PLA generals. However, Premier Li followed the tradition of the late Premier Zhou Enlai and the former Premier Wen in visiting the earthquake-affected regions immediately. If constitutional conventions exist in Chinese politics, they certainly embrace the premier's immediate inspections of the earthquake-plagued regions and the swift PLA mobilization after the earthquake's outbreak.

The Yaan earthquake showed the seismic authorities in the PRC needed more drastic improvement because the earthquake was detected 55 seconds after its occurrence while the pre-earthquake warning system simply failed (*The Sun*, April 22, 2013: A34). The National Seismic Bureau was criticized for failing to detect and predict the Yaan earthquake, not to mention warning the residents well before the earthquake occurrence. Some citizens on the Internet said that the seismic authorities even changed the magnitude of the earthquake from 5.9 to 7 after hearing the foreign assessment (*The Sun*, April 22, 2013: A34). Fortunately, nuclear facilities were not affected by the earthquake while water authorities quickly inspected the dams and lakes concerned (*Hong Kong Commercial Daily*, April 21, 2013: A1).

The CCP response was swift. On April 23, the CCP Politburo held a meeting in which President Xi Jinping led the Politburo members to observe minutes of silence to commemorate those people who sacrificed their lives in the Yaan earthquake, which occurred just two weeks before the fifth anniversary of the 2008 Wenchuan earthquake (*Ta Kung Pao*, April 24, 2013: A1). The Politburo listened to the report from the State Council's anti-earthquake and rescue command headquarters and decided that extra efforts would have to be made in several areas, including rescue operation, hygiene conditions, mass resettlement, repair work, and capital and logistical supplies. The Politburo decided that the Sichuan province had to consolidate its leadership, the State Council had to be responsible for division of labor and better coordination, and that the PLA and PAP had to maintain high morale so as to conduct the rescue operation, hygiene work, and resettlement work (*Ta Kung Pao*, April 24, 2013: A1). It also stressed that the entire CCP, nation and nationalities, had to maintain their cohesiveness in order to combat the earthquake—a language reminiscent of the official propaganda after the outbreak of the 2008 Wenchuan earthquake.

Donations from Hong Kong poured into the PRC. Albert Yeung's Emperor Group donated HK$1 million to the PRC for the relief operation, whereas the Hong Kong Red Cross also donated the same amount of money to the mainland (*Sing Tao Daily*, April 21, 2013: A5). The Taiwan Red Cross, after gaining the approval of the Interior Ministry, deployed 5 million yuan from the relief funds for the Wenchuan earthquake to be used to tackle the Yaan earthquake victims (*Ta Kung Pao*, April 22, 2013: A8). It also sent a rescue team to Yaan while many Taiwanese enterprises in the mainland, such as the Foxconn, decided to give donations and provide logistical supplies to the PRC. Although no single Taiwan person was hurt in the mainland, the Taiwan media paid special attention to the Yaan earthquake and was concerned about the plight of the victims. Each victim whose family member passed away due to the earthquake received 5,000 yuan of relief fund, while each parent whose only son or daughter died would receive 150 yuan per month as state subsidies (*Oriental Daily News*, April 28, 2013: A32).

There were reports saying that some counties affected by the Yaan earthquake exaggerated their economic losses so as to cheat outside donations, subsidies, and support. Three affected counties claimed that they suffered a huge economic loss of 169 billion yuan, a figure that was 21 times of the total GDP per capita of all the three counties in 2012 (*The Sun*, April 27, 2013: A26). Some citizens on the Internet questioned whether the three counties had a conspiracy in deceiving outsiders who would give more donations to support the disaster areas. Critics also raised a concern that the three local governments exaggerated their economic losses so as to secure more financial support from the provincial authorities for reconstruction work. Provincial-local politics came into play in the report of economic losses in the wake of the Yaan earthquake at the local level.

On the first anniversary of the Yaan and Lushan earthquake in 2014, the Sichuan authorities held a ceremony in memory of those who sacrificed their lives in the earthquake (*Hong Kong Commercial Daily*, April 21, 2014: A7). The national anthem was raised, accompanied by the national song, so that a strong sense of unity and nationalism was fostered. Reconstruction work was emphasized, including new urban and rural housing units as well as the repair work and reopening of four highways linking Lushan to the outside regions. The Yaan authorities announced that "not a single citizen felt cold and hungry" one year after the earthquake, a propaganda demonstrating their perseverance and success during the reconstruction process (*Hong Kong Commercial Daily*, April 21, 2014: A7).

On May 24, 2013, Yunnan province's Yingjiang county was affected by an earthquake of magnitude 5.6 and an aftershock with a stronger magnitude of 6.1 six days later, which altogether led to the collapse of 3,300 housing units, the damage to 18,000 other apartments, and the injuries of at least 39 citizens (*Sing* Pao, May 25, 2014: A19; The *Sun*, May 31, 2014: A21). On the morning of May 30, the seismic authorities gave 12 to 50 seconds of pre-earthquake warnings to 12 schools, where children were evacuated quickly (*Hong Kong Economic Journal*, May 31, 2014: A14). After the earthquakes occurred on May 24 and 30, one thousand PLA soldiers from Yunnan province were sent to rescue the affected victims in the scene (*Ming Pao*, May 31, 2014: A16). Altogether 41 aftershocks took place, injuring more citizens and prompting the seismic authorities to trigger level 2 of the emergency situation. An electricity station had its operation terminated by the earthquake, while rescue teams bringing temporary beds went to the affected county to help the earthquake victims immediately. Overall, 270,000 citizens in the county were affected, thus demonstrating the scale of the impacts of the earthquake (*Oriental Daily News*, May 25, 2014: A31). Judging from the reactions of the provincial authorities and the PLA, the PRC has already developed a kneejerk reaction to any eruption of earthquake. This natural response has demonstrated the policy learning on the part of the PRC government and seismic authorities in dealing with any earthquake similar to the 2008 Wenchuan earthquake.

MYANMAR'S HANDLING OF CYCLONE NARGIS IN MAY 2008 AS CONTRASTED WITH CHINA'S WENCHUAN EARTHQUAKE

On May 2, 2008, cyclone Nargis hit Myanmar or Burma, leaving at least 28,000 dead and 41,000 people missing. More than one million citizens were cut off from the supply of food, water, electricity, and telephones. Many Burmese were killed by a 3.5-meter-high tidal wave that crashed through the Irrawaddy delta at the peak of the cyclone (*The Globe and Mail*, May 7, 2008: A1). Schools were destroyed, survivors overwhelmed local hospitals, and rice fields were undermined. Rice prices doubled from 2,000 to 4,000 kyat (*Toronto Star*, May 13, 2008: A8). Reporters from foreign agencies said that the flooded areas could only be accessible by boats and helicopters could not even airlift logistical supplies to areas which were devastated by floods. Rangoon was also affected by the heavy flooding. Decaying bodies piled up the Irrawaddy river, leading to fears of an outbreak of contagious diseases.

The military junta in Myanmar waited for two days to accept international humanitarian assistance. The United Nations Children's Fund (UNICEF), the International Federation of Red Cross, and the Red Crescent Society were ready to help distribute tents, water purification tablets, and mosquito nets, but they were unable to get visas for their workers quickly (*The Globe and Mail*, May 7, 2008: A20). The United States offered to send the Disaster Aid Response Team but it did not receive any response from the Burmese government. The World Vision reacted to the natural disaster in Myanmar quickly, advertising in Canadian newspapers to receive public donations for the victims (*The Globe and Mail*, May 8, 2008: A13). Canada also offered help by sending a disaster team to Myanmar. The French foreign minister Bernard Koucher urged the United Nations (UN) Security Council to pass a resolution forcing the delivery of aid to Myanmar under the mandate of a responsibility to protect the Burmese people (*Toronto Star*, May 9, 2008: A1). Some 60 countries offered assistance to Myanmar shortly after the hit of the cyclone. The UN secretary-general Ban Ki-moon appealed to the Myanmar regime to allow foreign aid workers to go into the country, such as Amnesty International and Doctors Without Borders. The UN estimated that 1.5 million were suffering from the storm attack. On May 12, the United States delivered its first relief supplies to Myanmar (*The Toronto Sun*, May 13, 2008: 9). Nevertheless, Ban criticized Myanmar's military junta for being slow in response to the offer for help from the international community. The British government also criticized Myanmar for "making a natural disaster into a man-made disaster" (*World Journal*, May 18, 2008: A7).

The government of Myanmar was cautious toward international help, especially when the dissident political leader Aung Sang Suu Kyi, who was under house arrest and banned from holding public office, received the support of Western democracies like America and Britain. In May 2008, Aung criticized the military junta for claiming that 92 percent of the citizens accepted a new constitution and for asking the citizens amidst the cyclone disaster to cast their votes on constitutional amendments on May 10 (*World Journal*, May 18, 2008: A7). As a result of a strong xenophobic sentiment in the minds of the Myanmar ruling elites, who were suspicious of foreign intentions, they did not embrace the international assistance quickly at the early stage of the cyclone that hit the Irrawaddy delta.

The military junta was criticized by the international media for obstructing relief effort. The United Nations briefly suspended its aid shipments to Myanmar after the military was found to confiscate food supplies from two UN planes that landed on the Rangoon airport. The seized supplies included biscuits and rations for 95,000 people (*The Globe and Mail*, May 10, 2008: A19). The military in Myanmar also forced the victims of the May 3 cyclone to move from monasteries to tent camps, because the Buddhist monks organized funding and supplies for victims, who were seen by the ruling elites as being the supporters of the anti-military Buddhists (*Toronto Star*, May 16, 2008: A2). In September 2007, many Buddhist monks protested against the government for increasing fuel prices, leading to a military crackdown that led to the death of 31 people. Even rich Burmese people who wanted to donate rice or offer assistance to their country citizens were stopped by the military soldiers and officers, who told them that any aid had to be channeled through the military junta (*Toronto Star*, May 4, 2008: A3). Police also barred foreign aid workers from entering cyclone-affected areas, while some survivors were reportedly given spoiled or poor food rather than the biscuits sent by international donors (*The Toronto Sun*, May 14, 2008: 9). Internally and externally, the Myanmar regime exerted tight political control. The image of the Myanmar regime was so poor internationally that a foreign media described that "Burma flounders [over the cyclone] while China copes [with the Sichuan earthquake]" (*Toronto Star*, May 15, 2008: A1).

On May 19, Myanmar began a three-day mourning of those citizens who died in the cyclone attack. It was reported that at least 134,000 people died and were injured and 2.5 million people became victims and homeless (*Sing Tao Daily*, May 20, 2008: A13). The military junta allowed countries of the Association of Southeast Asian Nations to send health and relief teams into Myanmar but it still refused to give permission to foreign relief teams to do so freely. China also sent a 50-member health team to help Myanmar while Japan was invited to send observers to inspect the situation there. Thailand set up a giant food terminal in Rangoon so as to help Myanmar to conduct the relief operation (*The Toronto Sun*, May 25, 2008: 28). Obviously, Myanmar's military junta tended to trust the Asian neighbors far more than the Western countries.

Comparisons and contrasts can be made between the cases of the Myanmar cyclone and the Wenchuan earthquake. The Burmese military junta's reactions to foreign countries in 2008 were reminiscent of the PRC's resistance to foreign aid in the 1976 Tangshan earthquake, but the PRC officials tended to embrace foreign aid in the 2008 Wenchuan earthquake. The response of the Myanmar regime to foreign assistance was understandable, given the politics of Western support of the opposition leader Aung Sang Suu

Kyi. Wary of foreign support of Aung amidst the vote for constitutional amendments, the Myanmar regime distrusted the Western offers for assistance while agreeing to let Asian neighbors to help the country's relief efforts. As with the PRC government in 1976, the Myanmar regime tended to regard self-sufficiency and self-help as the most reliable way of tackling the cyclone disaster instead of accepting foreign aid without conditions. In a sense, the Maoist government in 1976 totally rejected foreign help, but at least the Myanmar regime in 2008 made a gesture to allow some international assistance amidst tremendous Western concerns and UN pressures.

In terms of the governmental capability, the PRC government in 2008 tended to be far more efficient and effective in dealing with the Wenchuan earthquake than the Myanmar regime in the same year. Internally, the Myanmar government had to deal with the potential opposition of Buddhist monks, but the PRC administration could work with most NGOs comfortably in partnership so as to conduct the rescue and relief operations. Although a minority of Chinese dissidents bemoaned the "tofu" buildings during the Wenchuan earthquake, they constituted a relatively weak social force in comparison with the politically active and potentially influential Buddhist monks in Myanmar.

THE 2011 TSUNAMI HITTING JAPAN AND CONTRASTS WITH THE 2008 WENCHUAN EARTHQUAKE

On March 11, 2011, Japan was hit by a tsunami catastrophe resulting from the Tohoku earthquake of magnitude 9, which took place at 125 kilometers east of Honshu and 380 kilometers from Tokyo. Half a million people had to be evacuated immediately and millions of them were deprived of drinking water, electricity supplies, and their homes. Even worse, the Fukushima nuclear power plant about 150 miles north of Tokyo suffered from the danger of a meltdown (Choate, 2011). The Japanese government was criticized by the members of the public and the political opposition for being unable to predict the earthquake beforehand and for being sluggish in its response to the natural disaster. Some people also criticized the Tokyo Electric Power Company—the owner of the nuclear power plant—for not providing sufficient information. However, volunteer groups and NGOs quickly offered assistance to help the tsunami victims and survivors. Despite the volunteerism shown in the Japanese society, the mega-earthquake and its resultant tsunami in 2011 made the 1995 Kobe earthquake of magnitude 7.9, which killed 6,400 people, and the 2005 Hurricane Katrina, which led to a death toll of 1,800 in New Orleans, pale in comparison (Choate, 2011). The scale of the natural disaster in Japan in March 2011 was relatively much larger: a strong mega-earthquake followed by a lethal tsunami and then a looming nuclear crisis. Even though Japan implemented its building codes strictly to minimize the damage of any earthquake, the 2011 tsunami was so powerful that the 30-feet high waves destroyed many buildings and claimed many lives. Although Japan's power plants were protected by seawalls as high as 25 feet, the high and strong waves resulted from the tsunami simply made the Japanese defense useless.

The responses of the Japanese government were already swift, mobilizing the Self-Defence Force to participate in the rescue and relief operation immediately. In the 1995 Kobe earthquake, Japan had denied the international offers for assistance on the grounds of language and communication difficulties (Choate, 2011). However, the 2011 tragedy made the Japanese government welcome foreign help as the globalization of information, communication, and intergovernmental and nongovernmental assistance became inevitable. Moreover, the Japanese government learned from the 1995 Kobe earthquake by developing an early-warning system.

Although Japan's governance was regarded as relatively "satisfactory" among the industrialized countries, it was far from setting up a "model" for other parts of the world, especially in its lack of transparency in dealing with the earthquake tsunami in 2011 (Kaufmann and Peniciakova, 2011). Minutes prior to the 2011 earthquake tsunami, alerts were sent to millions of cell phones, railway trains were terminated, and stations interrupted so as to broadcast warnings (Kaufmann and Penciakova, 2011). The Japanese government also responded quickly to the tsunami through the mobilization of 100,000 troops to rescue and search the survivors. However, in terms of its response to the crisis at the Fukushima nuclear power plants, which experienced explosions in reactors 1 and 3, there were "serious shortcomings in leadership and governance" (Kaufmann and Peniakova, 2011). Radiation levels increased and penetrated in food and drinking water in areas far from the evacuated zones. Although workers and soldiers prevented a nuclear meltdown by flooding reactors with seawater, the crisis exposed the governing clumsiness and administrative incompetence of the Tokyo Electric Power Company that had reportedly covered up inspection data showing cracks in some reactors in 2003, refused to answer questions from the press shortly after the 2011 tsunami, and reacted quite slowly to the crisis by shutting down its last operating nuclear reactor in the country for repair maintenance on March 26, 2012 (Kaufmann and Peniciakova, 2011). Lax supervision in the safety of nuclear power plants turned out to be costly in time of crises, as the Nuclear and Industrial Safety Agency under the Ministry of Economy, Trade and Industry had already found in 1977 that seven out of 12 public utilities in Japan falsified their safety records. Regulatory gaps in Japan accounted for the eruption of the nuclear power plant crisis immediately after the 2011 tsunami attack.

Akira Nakamura and Masao Kikuchi (Nakamura and Kikuchi, December 2011: 893–899) have argued that the Fukushima nuclear fiasco in 2011 demonstrated a whole range of problems in Japan's public administration, including the lax supervision of nuclear facilities on the part of the public and private authorities, the lack of central leadership, the persistence of electric power industry monopoly without sufficient checks and balances, and the existence of a fragmented policy-making structure in which several agencies had control over the development of nuclear resources, energy, and regulations but did not shoulder any responsibility in times of crises (Nakamura and Kikuchi, December 2011: 897). Nuclear plants were rapidly developing in Japan without the proper policy-making apparatus that could monitor the entire process. In short, the concept of "critical infrastructure protection" was either weak or absent (Nakamura and Kikuchi, December 2011: 898). Japan has to learn a bitter lesson from the Fukushima nuclear tragedy in March 2011.

Zare and Afrouz have analyzed the implications of the 2011 tsunami for Japan's crisis management one year after the earthquake, which caused the death of 15,854 people, 3,167 missing and 26,992 injuries (Zare and Afrouz, 2012: 12). They have observed the "social capital" facilitating the calm behavior of citizens and NGOs following the tsunami, demonstrating the Japanese people's high respect of their roles, solidarity, and moral values without showing any chaos in acquiring aid or any motivation in looting (Zare and Afrouz, 2012: 12–13). The government reacted promptly by holding meetings of the National Committee for Emergency Management chaired by the prime minister, declaring a state of nuclear emergency in light of the situation in the Fukushima nuclear power plants, mobilizing the ministries of agriculture and fisheries and finance to trigger the rescue operations, and utilizing all the main highways north of Japan for the emergency response to the tsunami.

Yet, the tsunami had serious consequences. Radioactive pollutions turned out to be the most harmful repercussion on residents (Zare and Afrouz, 2012). Many suffered from psychological trauma and stress, while the tsunami wreaked havoc with Japan's jungles and agricultural production. Wood demand and imports increased sharply as wooden houses are popular in Japan. Women affected by the tsunami had to find and switch occupations as the government created 20,000 jobs within a month after the disaster, but still the jobs created were insufficient.

Zare and Afrouz concluded that the Japanese people displayed impressive calm and orderly behavior after the tsunami—a model for citizens in those developing countries affected by earthquakes—and that seismologists in the world should revise their models of estimating and predicting the time, magnitude, and occurrence of earthquakes. They noted the management of nuclear power plants as a "blind spot" in Japan's disaster management, and the economic priority over morality in the Japanese authorities who continued to export nuclear technology to other countries (Zare and Afrouz, 2012: 19–20).

As a matter of fact, the calm responses of the Japanese people and of the local governments after the 2011 tsunami were not surprising, for Japan had learned a lesson from the earlier 1995 Kobe earthquake, which revealed the urgent need for the local governments to provide the service of free building seismic evaluation of wooden houses to promote seismic retrofit, and the need for the citizens to collaborate with firefighters, schools, companies, and hospitals to conduct pre-earthquake drills (Saito, 2007). The Japanese Diet passed an Act for Promotion of the Earthquake Proof Retrofitting of Buildings in December 1995 to make many buildings more resistant to earthquakes than ever before (Suganuma, 2006: 99). Although ordinary homes were not included in the implementation of the act, the earthquake consciousness of Japanese citizens was deeply entrenched in the society. With almost two decades of persistent education efforts at preparing citizens for earthquake attack, Japan had nurtured a relatively mature civil society in coping with the eruption of any natural disaster.

The mobilization of the Japanese Self-Defence Force (SDF) was noteworthy in the 2011 tsunami in Japan. Within 72 hours of the earthquake, the Ministry of Defense mobilized the SDF and its reserves. The SDF established a Joint Task Force to coordinate the relief efforts by ground, air, and maritime forces, with Lt. General Eiji Kimizuka as the commander and General Ryoichi Oriki as the chief of staff of the joint staff (Tatsumi, November 2012: 3). At the apex of the rescue and relief operations, 106,000 personnel were mobilized to provide support for rescue and recovery, transport assistance, livelihood assistance, and emergency rehabilitation. The SDF also helped cool the reactors and spent fuels, decontaminated the region around Fukushima, and monitored the radiation level and situation there.

Apart from the SDF, the United States military under the leadership of commander Lt. General Burton Field, and later under the Pacific Fleet commander Admiral Patrick M. Walsh, conducted the Operation Tomodachi that involved 24 naval ships, 189 aircraft, and 24,000 personnel (Tatsumi, 2012: 3–4). The US Marines that were trained for chemical, biological, radioactive, and nuclear response played a crucial role in gathering the radiation intelligence in the development of the Fukushima Daiichi Nuclear Power Station. The United States Nuclear Regulatory Commission, Department of Energy, and the Department of Defense contributed assistance to help the Japanese government to deal with the Fukushima nuclear power plants crisis, including the utilization of on-the-ground expertise, decontamination of assets, the supervision of contamination of food and water, aerial detection, the deployment of high-pressure water pumps and fire trucks, and the use of protective gear from radioactivity (Fierckert and Chanlett-Avery, 2011: 4–5). The American Navy also provided two water barges of 500,000 gallons of fresh water for the cooling purposes in the damaged nuclear reactors, while the Marines' Chemical and Biological Incident Response Force was deployed from the United States to train the Japanese SDF. Both the SDF and United States military set up coordination cells known as Bilateral Coordination Action Teams in Ichigaya, Yokota, and Sendai. For the first time in Japan's postwar history, the SDF was fully mobilized to deal with a crisis involving nuclear power plants undermined by a tsunami, and cooperated with the American military during a state emergency. The lessons learned for the Japanese defense capabilities include, according to Yuki Tatsumi, firstly the SDF's heavy reliance on the private sector for command and control, communications, computers, intelligence, surveillance, and reconnaissance, thus raising the question of whether the communication system could survive during emergency. Second, this reliance on the private-sector technology hampered the SDF's rescue and relief operations. Third, the SDF's transport and logistics capabilities were also inadequate to sustain long-term mobilization, implying that it would have to forge closer partnerships with private-sector transport companies in order to gain better access to vehicles in future emergency scenarios. Fourth, given Japan's previous policy of keeping its nuclear power plants for civilian use only, the Ministry of Defence was not seen as one of the stakeholders in nuclear power safety but it would have to be incorporated in any nuclear safety regime in the future (Tatsumi, 2012: 4). Fifth, institutionally, the SDF would have to strengthen its internal reorganization, including the clarifications of the responsibilities of the chief of staff of joint staff, and the need to deal with officers who suffered from the post-traumatic stress disorders during the rescue and relief operations.

Alongside with the Self-Defence Force, the Japanese police played a crucial role in the rescue and relief operation after the 2011 tsunami. The National Police Agency in July 2012 released a white book that contained a section discussing in detail its action as the earthquake tsunami devastated the Tohoku coastal region on March 11, 2011 (*Japan Times*, August 4, 2012). Among the 16,000 people killed in the natural disaster, they included 25 police officers and five others still missing. On March 11, when wireless communications for disaster prevention did not work properly, the police in the Nankai ocean trough subduction zone sent accurate instructions to local residents and evacuated them safely, including some 3,750 people in Iwate, Miyagi, and Fukushima prefectures. They searched for missing people and were joined by emergency aid units composed of police from various parts of Japan and the 6,400-member rapid reaction units. In response to the shortage of police officers, the National Police Agency in 2012 increased the size of the rapid reaction units to some 10,000 officers.

After the 2011 tsunami, the Japanese government reflected on the natural disaster and admitted that pre-earthquake disaster assumptions would have to be reexamined, meaning that tsunami resulted from earthquake and the height and intensity of such tsunami would have to be reassessed and predicted in the future (Government of Japan, *Lessons Learnt from the Great East Japan Earthquake*, 2011: 195). It also admitted that previous hazard maps and coastal protection facilities were not enough to deter the unprecedented high level of the tsunami waves that caused much destruction to the interior regions of Japan. The Japanese government recognized the need to conduct scientific study of ancient documents, historical materials, and surveys of tsunami deposits as well as ocean topography so as to undertake long-term research on seismic movements. Indeed, forecasting earthquakes is difficult, as acknowledged by the Japanese government, which also stated that there would be a need to create resilient communities involving women participation, better land use planning, and better designs of evacuation buildings and routes in the future (Government of Japan, *Lessons Learnt from the Great East Japan Earthquake*, 2011: 200–201). Resiliently, the Japanese government concluded in the following way:

> We cannot simply mourn the losses and damage inflicted by the Great East Japan Earthquake. While mourning the damage, we must stand and face whatever catastrophe disaster may occur and do all that we can to protect human life. On the basis of this conviction, we must further strengthen earthquake and tsunami countermeasures while carrying through the recovery and reconstruction of the devastated areas. We must systematically build a nation and cities that are even stronger against earthquakes and tsunamis as we educate and raise the awareness of the public about earthquake and tsunami disasters. Moreover, there is a need for disaster-management related fields of science and engineering, humanities and social science, and life sciences to join forces in a collective effort to carry out scientific surveys examining the reasons for the occurrence of disasters and comparing local characteristics, such as whether or not damage was inflicted, as well as to strengthen the research system (Government of Japan, *Lessons Learnt from the Great East Japan Earthquake*, 2011: 203–204).

The fact that it recognized the need for comprehensive disaster management measures was a progressive step forward, demonstrating policy learning from the experiences of the 2011 earthquake and tsunami.

Comparatively speaking, the 2011 tsunami hitting Japan had commonalities and differences with the 2008 Wenchuan earthquake in China. First and foremost, both the governments of Japan and China did not anticipate and pre-warn the occurrence of the large-scale natural disasters. Although the Japanese government had learned from the 1995 Kobe earthquake and improved the pre-earthquake emergency preparation, still it could not accurately predict the occurrence of the 2011 earthquake tsunami. Studying the PRC governmental responses to the Wenchuan earthquake, management experts David Olson and Desheng Wu have argued that China should develop better mechanisms for earthquake risk management: one better predicting earthquake damage magnitude and the other preparing response team performance (Olson and Wu, 2010: 478). They concluded that pre-planning in the aspects of gathering and organizing data, developing action plans at the national, regional, and local levels, and organizing trained people into teams to respond to natural disasters with all the necessary rescue, medical, food, and transport systems would be necessary for the PRC to cope with any natural calamities in the future (Olson and Wu, 2010: 492). Some mainland Chinese public health experts have also agreed that emergency response plans were lacking at the local level during the Wenchuan earthquake (You, Chen and Yao, 2009: 379–394). They called for the need on the part of the PRC government to formulate emergency plans for its staff and the public, to encourage citizens to be self-reliant in time of crisis, to train emergency workers in conducting rescue operations, to improve cross-sectors cooperation, and to organize social groups in a better way for the sake of coping with any natural disaster in the future. Indeed, more can be done in both Japan and China in terms of anticipating natural disaster, sending pre-disaster warnings, and dealing with other side effects, such as the nuclear power plants crisis in Fukushima.

Another commonality in the responses of Japan and China to the 2011 tsunami and 2008 Wenchuan earthquake was the resilience of the Japanese and the Chinese people. While the Japanese displayed maturity, calmness, and their obedience of law and order in the days after the tsunami attack, the majority of the Chinese citizens also demonstrated resilience. The maturity of the Japanese civil society, where volunteer groups and individuals acted and organized themselves immediately, was perhaps even more advanced than the Chinese civil society, which remained dependent on the Chinese Party-state for guidance and approval. However, if the Japanese and Chinese were in general law-abiding citizens, their resilience did constitute one of the most important factors contributing to the societal success in forging a partnership with the government in resisting the natural disasters.

In terms of the maturity of citizens, the Japanese appeared to be far more mature and much calmer than the Chinese in dealing with natural disasters. The orderly nature of the Japanese citizens during the process of evacuation and waiting for relief and their law-abiding tendency were the hallmarks of the Japanese public reactions to the 2011 tsunami, perhaps unlike China where a minority of citizens violated the law and some corrupt officials were criticized for embezzling the logistical supplies. This phenomenon reflected the governing environment of Japan and China, with the former being internationally rated by Transparency International as relatively "more clean" than the latter. After all, anti-corruption drive in the PRC has been encountering huge obstacles and resistance since the Xi Jinping era, not to mention the Hu Jintao era when the Wenchuan earthquake broke out.

With regard to the most prominent difference, while Japan embraced international aid from foreign countries after the 2011 tsunami attack, the special assistance acquired from the American military in response to the nuclear power plants crisis was unique, distinguishing the Japanese government's response from the PRC's own reliance on the PLA to tackle the rescue and relief operations after the Wenchuan earthquake. As with Japan, the PRC also embraced international relief and donations. But the technical and advanced military support from the United States military for Japan after the 2011 tsunami attack demonstrated the deep American-Japanese ties in making a joint effort at reducing the risks and harms resulting from the unprecedented tsunami and nuclear power reactors crisis.

The nuclear power plants crisis stemming from the earthquake and tsunami in Japan in March 2011 was a far cry from the 2008 Wenchuan earthquake. At the very least, the Wenchuan earthquake did not bring about other serious side effects, albeit landslides, mudslides, dam cracks, and landscape alterations did take place. Fortunately, the PRC's nuclear power plants were not affected, thus limiting the scope of any secondary natural disaster resulting from the Wenchuan earthquake. In Japan, the danger of a secondary natural disaster, namely nuclear meltdown, was so immense that the SDF would have to require the assistance of the United States military to prevent the potential crisis from being exacerbated further.

THE CHILE EARTHQUAKE IN 2010 AND 2014

On February 27, 2010, an earthquake of magnitude 8.8 hit Chile, killing 521 people, and creating a tsunami that eventually hit as far as Japan's Tohoku's region. The Chilean response was initially chaotic but it improved quickly, including the mobilization of the military to control any looting on the streets and to conduct rescue, debris removal, and reconstruction operations (Hinrichs, Jones, Stanley and Kleiner, 2011: 5). Some observers have attributed the relatively low death toll to the human rights-based framework that was deeply entrenched in the Chilean society. For instance, the Chile Grows with You was an organization established in 2007 providing educational, mental, health, psychological, social support for children, who could build up their networks and capability to cope with any natural disaster (Arbour, Murray, Arriet, Moraga and Vega, 2011: 62–73). Immediately after the 2010 earthquake, the Chile Grows with You could provide medical and practical support for not only the children but also pregnant women. The protection of children's rights is arguably one of the unique hallmarks of the Chilean approach in dealing with the eruption of any earthquake.

On April 1, 2014, a powerful earthquake of magnitude 8.2 struck off northern Chile, triggering a small tsunami with a 2-meter height and killing six citizens (Henao, 2014; Esposito and O'Brien, 2014). In the city of Iquique, about 300 inmates escaped from a female prison and the military was sent to help the police to maintain law and order there. Later 130 escapees identified themselves voluntarily to the police, whereas 26 others were arrested (*Hong Kong Sky Post*, April 3, 2014: 22). In the city of Arica, houses were destroyed and electricity was cut off. The strongest earthquake in Chile's history took place in 1960, when an earthquake of magnitude 9.5 hit the country and killed some 5,000 people. Another strong earthquake happened in 2010, when 500 residents died and 220,000 homes were destroyed.

The relatively small death toll in Chile in April 2014 was attributable to the strict building codes that helped thousands of homes and buildings sustain any structural damage. Apart from strict building codes that were enforced in Chile even before the 2010 earthquake that killed 500 people, Chile conducted simulation exercises to prepare citizens for the occurrence of earthquakes. In May 2012, 500,000 people participated in a drill in Valparaiso (Sanchez, 2014). Moreover, once the Chilean government alerted the people of the forthcoming tsunami in April 2014, many citizens were evacuated at the Antofagasta city; some stayed at the top floor of the building in the city of Iquique during a vertical process of evacuation; and others walked to the higher ground in the Talcahuano city (Saavedra, 2014).

The rights-based approach and the strict building codes in the Chilean experiences can be learned by the PRC. Given the fact that the PRC government is more politically paternalistic in its approach to coping with human rights issues, the rights-based emphasis cannot be entirely applicable to the Chinese setting. Nevertheless, the social support for children can be consolidated in the PRC, thus providing a safety valve in the protection of children during any natural disaster. In fact, given the assertiveness of nongovernmental organizations in China after the Wenchuan earthquake, it can also be argued that a civil society-based approach in dealing with earthquakes has been gradually and mildly developed in the PRC since 2008. Indeed, the civil society has to work in partnership with the Chinese state and the state remains dominant in the process of rescue, reconstruction, and legitimization. Moreover, the strict building codes that have been observed in Chile have positive learning implication for China, where many buildings were developed without observing building codes in preparation for any earthquake occurrence. In short, the Chilean experiences can be selectively transplanted into the PRC circumstances in preparation for any earthquake in the future.

CONCLUSION

This chapter uses different case studies to compare and contrast with the PRC handling of the Wenchuan earthquake in 2008, such as the Tangshan earthquake inside the PRC in 1976, the outbreak of SARS in 2003, the havoc wreaked by cyclone Nargis in May 2008, the 2011 earthquake and its tsunami in Japan, and the recent Chilean experiences. Internally speaking, the PRC made tremendous progress in managing crises, as evidenced in the relatively weak responses of the government to the Tangshan earthquake, the belated but more open response of the Beijing administration to the seriousness of SARS in March 2003, and the immediate mobilization of provinces and mass media to tackle the Wenchuan earthquake in 2011. Undoubtedly, the PRC government did learn from the series of crises and improved its crisis management methods (Zhao, 2009). This policy learning could be seen in the PRC government's handling of the Yushu earthquake in 2010 and the Yaan earthquake in 2013. Compared with the way in which the socialist government of Myanmar handled the impacts of cyclone Nargis in May 2008, the PRC government performed far better than Myanmar in many aspects, including the attitudes toward foreign help and aid, the internal rescue operations, the partnership with nongovernmental groups, the regime legitimacy in the mind of most citizens, and the international image in the eyes of foreign countries. Compared with the Japanese government in handling the 2011 tsunami, the PRC was perhaps more fortunate to shun a scenario of having to deal with any crisis in its nuclear power plant, but both countries witnessed active citizen participation in the rescue and relief operations and they demonstrated the persistence of a civil society supportive of the government in time of national emergency.

Conclusion

This book argues that internally and historically the PRC government has made significant improvements in managing crises since the 1976 Tangshan earthquake, which demonstrated the difficulties of the rescue operations and which killed 240,000 people. Although scholars like Col and Chu have argued that the Qinglong county was prepared days before the earthquake and represented an example of the "best practice in public administration" (Col and Chu, 2001: 584), the tremendous death toll appeared to demonstrate the validity of the reverse phenomenon. The PRC in 1976 was in general unprepared for the Tangshan earthquake. However, policy learning from crises took place in mainland China, where the outbreak of the SARS in 2003 demonstrated local cover-up in the first stage of the crisis but forced the government to be more open, transparent, and candid in dealing with the mysterious viruses at the central, provincial, and local levels since the public acknowledgment of the severity of the disease by President Hu in April 2003. The Wenchuan earthquake in 2008 was a further testimony to the improvement of the PRC responses to the eruption of natural disaster. The full mobilization of the provinces and cities to participate in the rescue and relief operation, accompanied by the utilization of the official mass media to publicize the tragedy to all the Chinese citizens in the mainland, became the tools through which the central government in Beijing dealt with the earthquake crisis in an effective and a prompt manner. In short, the PRC government made tremendous strides in crisis management from 1976 to 2008. The series of earthquakes in the PRC after 2008 have demonstrated that the PRC responses have displayed a consistent pattern of swiftness, including the rapid mobilization of the PLA, the premier's immediate visits, the partnership between the state and the civil society, the persistence in reconstruction work, and most importantly the success in legitimizing both the CCP and the state throughout the entire processes of rescue and reconstruction.

In terms of regime legitimacy, it remained unchallenged by the very weak or perhaps almost nonexistent civil society in the PRC in 1976 as the government adopted a top-down and relatively hard-line approach in ruling the country. Despite the heavy death toll, the mobilization of the PLA and police in the rescue and relief operations after the Tangshan earthquake demonstrated the ease in which the central government governed the country in a top-down fashion. In the early half of 2003, the eruption of SARS did not really challenge the legitimacy of the PRC government, which eventually responded to the virus quickly and effectively after April. Most importantly, the 2008 Wenchuan earthquake even provided a golden opportunity for the PRC regime to consolidate its own legitimacy through provincial, media, and public mobilization as well as the rapid responses from the leaders of the CCP-state. By utilizing foreign assistance and sympathy, the PRC's regime legitimacy remained secure, solid, and relatively high in 2008 during which the Summer Olympics was expected to push its legitimacy to an apex. The holding of the Olympics represented the glory of the Chinese nation-state, exerting tremendous pressure on the central government in Beijing to deal with the rescue and relief operations after the Wenchuan earthquake decisively, promptly and effectively. If the Tiananmen incident on June 4, 1989 marked a blemish in the history of the PLA, which was deployed to suppress the student democrats, the 2008 Wenchuan earthquake arguably provided a golden opportunity for the PLA to re-legitimize itself in the eyes of the mainland Chinese citizens. The redemption of legitimacy was arguably the most significant impact of the Wenchuan earthquake on the PLA even though it showed some degree of difficulties in the rescue operations.

The rapid responses of the CCP-state deserve scholarly attention as its leadership in managing the Wenchuan earthquake consolidated the entire regime legitimacy. Nevertheless, Premier Wen's criticism of the PLA in the early days of the post-earthquake rescue and relief operations demonstrated the internal opinion differences of the CCP and military leadership. As one of the top CCP-state leaders, Wen was eager to witness a faster mobilization of the PLA to rescue the earthquake victims and survivors, but some PLA leaders believed that the severity of the weather conditions complicated and hindered the rescue operations. In spite of the opinion differences between Wen and some PLA leaders, the military displayed effectiveness in the entire rescue operation even though it also demonstrated some technical difficulties. Overall, however, the modernization of the PLA was tested successfully by the rescue operations. In the minds of most citizens, the PLA rescue success contributed to the consolidation of the CCP legitimacy and the realization of the people's principle of the army. If political power grew out of the barrel of the gun, as the late CCP Chairman Mao Zedong remarked, then the CCP legitimacy in the post-Wenchuan rescue and relief work was undoubtedly strengthened by the mobilization of the PLA forces.

The role of Premier Wen, as argued before, was different from the late Premier Zhou Enlai in one important aspect. While Wen was a civilian leader urging the PLA to be more efficient in its rescue operation shortly after the Wenchuan earthquake, his stature was actually no match with Premier Zhou who had long been both a civilian-military leader. Unlike Zhou who commanded tremendous respect and high status in the eyes of the PLA generals and officers, and who praised the PLA work in the 1966 Xingtai earthquake, Wen adopted perhaps a more impatient approach in criticizing the PLA. As a result, it was natural that the PLA's high command was displeased with Wen's behavior. Still, the premier's visit to the earthquake-affected regions has become a constitutional convention in Chinese politics, for it symbolizes the necessity of the Party-state to respond to natural disasters promptly, thus winning the hearts and minds of the ordinary people.

The consolidation of the legitimacy of the CCP-state was also facilitated by the emergent social groups, particularly NGOs and volunteers in support of the state-led and military-led rescue efforts. NGOs that loomed in the post-earthquake relief forged a partnership with the CCP-state rather than having any confrontational relations. If those NGOs had confrontational relations vis-à-vis the government at the central, provincial, and local levels, they referred to a minority of parents, who mourned the deaths of their beloved children and who put the blame of the "tofu" buildings on the corrupt officials. As the PRC remains to be governed by a paternalistic socialist regime with tight control on citizen protests, any political dissidents stemming from the Wenchuan earthquake were bound to be curbed and suppressed.

As a matter of fact, since the global rise of the PRC in the early 2000s, the concept of national security in China has been broadened to encompass crisis management, including disaster management. Mainland Chinese scholars have also defined the concept of national security as including various dimensions: food security, economic and monetary stability, political stability, scientific advancement and competition for talents, cultural protection, information security, national defense and military security, environmental and ecological security, societal stability, anti-terrorism, and international security (Cao, 2008: 315–332). Ecological security embraces the ways in which the PRC government should protect the environment, take precautionary measure against natural disasters, and engage in research that would mitigate the impacts of environmental and climate change as well as natural disasters. In this aspect, earthquake crisis management belongs to the dimension of ecological security. Specifically, earthquake and its related side effects on any natural disaster would endanger China's national security, just like the 2001 earthquake that produced an unprecedented tsunami wave in Japan. Exactly because the PRC government has seen natural disasters as national security issues, it took prompt measures to tackle the 2008 Wenchuan earthquake decisively and effectively.

The scale of the natural disasters also determines whether the governmental responses can bolster, rescue, or revive its regime legitimacy. The Wenchuan earthquake was of such a large magnitude that it wreaked havoc with many regions. It really tested the capability of Beijing, the Sichuan provincial government, and the local government to tackle the impacts and legacies. What was fortunate in the Wenchuan earthquake was its inability to undermine any nuclear power plants and facilities in China, unlike the height of the tsunami waves that inundated the Fukushima nuclear power reactors in Japan in March 2011. Hence, the impacts of the Wenchuan earthquake were comparatively less widespread than that of the 2011 earthquake in Japan, which suffered immediately from the powerful tsunami waves sweeping across Fukushima.

Compared to the 2010 earthquake that killed 230,000 people in Haiti, the Wenchuan earthquake appeared to be less lethal in terms of scale and damage. When an earthquake of magnitude 7 hit Haiti on January 10, 2010, the Haitian government was totally unprepared—a situation exacerbated by the fact that the rescue operation was difficult because of the terrain. In many regions, there was only one road connecting various hills and houses built on either sides of the road that was devastated and blocked by the earthquake, making it extremely difficult for rescue teams and tools to reach the earthquake-hit zones (Bapat, 2010: 1287). Although the Wenchuan terrain was fully of valleys and rivers that created similar difficulties for the rescue teams to conduct their operations, the mobilization of the PLA in the rescue operation appeared to mitigate the earthquake's detrimental impacts, unlike Haiti where the government was almost totally decapitated by the 2010 earthquake, the national palace was destroyed, and the United States army, navy and air force had to be requested and deployed immediately to rescue the Haitians.

If the Qinglong county's pre-warning systems worked well days before the 1976 Tangshan earthquake, as Col and Chu found, its experiences could not be repeated easily in most other affected regions in the Tangshan earthquake and also in the 2008 Wenchuan earthquake. According to Col and Chu, the Qinglong's responses were marked by three main characteristics. First, public administrators in the State Council took various initiatives, including the formulation of Document Number 69, which told seven municipalities and provinces in Northeast China that a major earthquake would be likely and which urged them to take precautionary measures. Second, the Qinglong county was punctuated by an increase in community awareness of the possible earthquake through widespread and intensive public campaigns (Col and Chu, 2001: 587). Third, with regard to the role of earth scientists, two weeks prior to the earthquake, they had shared geological information and discussed geo-electric, seismic, meteorological and hydrologic data at a conference held in Tangshan. But "the lack of familiarity with some important aspects of precursory signals (e.g., amplitude vs. length of time of signal) resulted in earthquake predictions of M5" (Col and Chu, 2001: 590). The State Seismological Bureau's prediction of the magnitude of the Tangshan earthquake was less accurate than the scientists working at the Qinglong county, for the latter anticipated an earthquake with a magnitude of 6 to 8.

The three main features in the Qinglong county could not be seen in all other regions in Tangshan, according to the eyewitness account from Qian Gang, whose work depicted a different picture in which all the regions were unprepared, despite the fact that the behavior of animals looked bizarre days before the earthquake, just like the situation of toads and frogs running out from their hiding areas days prior to the 2008 Wenchuan earthquake (Qian, 2010). In fact, most scientists and seismological officials could not really anticipate the Tangshan and Wenchuan earthquakes before their outbreaks. Even if the Qinglong county's scientists could do so, officials in other regions in Tangshan in 1976 could not take immediate measures to save the lives of the citizens. Hence, the findings of Col and Chu could not be generalized to other regions in Tangshan and their arguments need to be reassessed with great caution and reservations, especially in light of the inability of most seismological authorities in the PRC to predict the 1976 and 2008 earthquakes while the pre-earthquake behavior of animals ironically did so.

However, Col and Chu have offered some insights on the interactions between public administrators, citizens, and scientists on the one hand and their level of participation in mitigating the impacts of earthquakes on the other hand. On preparedness, they emphasized the importance of the networking of all levels of society and the open communication of disaster information (Col and Chu, 2001: 594). With the benefit of hindsight, the 2008 Wenchuan earthquake and its aftershocks did reveal the extent of societal networking and the sharing of information through mass media reports. These societal networks were expressed in the formation and prompt action of volunteer groups and NGOs, whereas the media reports did bring about an unprecedented openness in tackling the rescue and relief operations effectively. Arguably, societal networking and open communications represented a progressive step forward in the public and governmental responses, respectively, to the Wenchuan earthquake in 2008.

Table C.1 below demonstrates the circumstances of the 2008 Wenchuan earthquake by adapting the analytical framework advanced by Col and Chu, who perceived the natural disaster's outcomes as a result of the dynamic relations between citizens, public administrators, and scientists. Col and Chu have not considered regime legitimacy as part and parcel of the natural disaster's outcomes. Adapting the analytical framework of Col and Chu, the introduction has included the concept of regime legitimacy into our study.

Table C.1 The Participation of Citizens, Public Administrators and Scientists and the Outcomes of Natural Disasters in the 2008 Wenchuan Earthquake, 2008 Myanmar Cyclone, and 2011 Japanese Earthquake Tsunami

Participation	*Citizens*	*Public Administrators*	*Scientists*	*Natural Disaster Outcomes*
Unaware, little or no action	Ignorant, unaware	Too busy with daily affairs to consider any natural disaster	Do not think about how natural disasters affect people	Total tragedy, serious crisis of **regime legitimacy**
Example above: Myanmar's Cyclone Nargis in May 2008 as it killed 28,000 people and plunged the Burmese regime into a crisis of legitimacy				
Educated, Moderate	Aware, take precautionary measures, drills were made to cope with earthquake	Gather information, formulate plans, assess relief needs, but cannot evaluate the accuracy of predictions	Cannot make accurate predictions, but produce predictions with a degree of probability	Tragedy, massive relief, having impacts on other natural disasters, such as nuclear power plants, **regime legitimacy** was affected negatively
Example above: The 2011 earthquake tsunami made the Japanese government suffer from a legitimacy crisis due to its sluggish responses, especially the Fukushima nuclear power plants crisis resulting from the tsunami.				
Unaware, low level	Citizen's lack of education, preparation, and drills	Lacking education of the public, but mobilizing the media, NGOs, provinces, and police	Precursor signals were absent except for the behavior of animals	No early warning, immediate evacuations of communities, minimize deaths and rescue survivors, winning the hearts and minds of the people, and thereby **regime legitimacy** can be protected and consolidated
Example above: The 2008 Wenchuan earthquake did not impact negatively on the regime legitimacy in the PRC which successfully				

If the participation of the members of the public, officials, and scientists can be maximized, the impacts on regime legitimacy would also be positively enhanced. In a nutshell, the lack of crisis preparation can plunge a regime into a serious crisis of legitimacy in the event of an outbreak of the earthquake. A moderate level of citizen, official, and scientist participation would mitigate the negative impacts of earthquake occurrence on the regime legitimacy, which also depends on the measures taken by the government concerned. If the government can react swiftly to earthquake outbreak by taking prompt relief measures, rescuing citizens effectively, and implementing reconstruction work rapidly, regime legitimacy can be consolidated. On the contrary, sluggish governmental responses to earthquake would de-legitimize the regime. Finally, if the level of participation on the part of citizens, officials and scientists is high, then the impacts on regime legitimacy are arguably positive.

As shown in table C.1, while the cyclone Nargis hitting Myanmar in early May 2008 was characterized by unprepared citizens, inefficient public administrators, and the inability of scientists to predict its occurrence, the natural disaster's outcome was catastrophic, leading to serious death tolls and the regime's legitimacy crisis. The 2011 earthquake tsunami plunged the Japanese government into a legitimacy crisis due to its sluggish responses, especially the Fukushima nuclear power plants crisis resulted from the tsunami. Although the Japanese citizens were educated, prepared, and aware of the dangers produced by earthquakes, scientists were unable to predict the earthquake and tsunami's occurrence. As such, the responses of the Japanese government became critical to its legitimacy. Although the Japanese government was criticized for its clumsy responses to the Fukushima nuclear power plants crisis, the American military came to the rescue, preventing the plants from having a meltdown and thus averting the crisis. In a sense, the legitimacy of the Japanese government was propped up by the external superpower immediately after the 2011 tsunami that triggered the crisis of the nuclear power reactors. The Wenchuan earthquake, however, was punctuated by unprepared citizens, inability of scientists to anticipate the natural disaster, but by an active citizenry which stimulated the growth of a civil society and volunteerism in assisting the government in its rescue and relief operations. Most significantly, the full mobilization of provinces, cities, mass media, the PLA, the police, and citizens and groups by the central government in Beijing turned out to be a huge success. Political mobilization could, in short, avert a crisis of legitimacy due to the occurrence of the Wenchuan earthquake. Reversely speaking, the Wenchuan earthquake even created a golden opportunity for the PRC regime or the Party-state to bolster its legitimacy with less than three months away from the opening of the Summer Olympics in Beijing.

Col has also emphasized the importance of the responses from the subnational and local governments to natural disasters in different stages: (1) preparedness in which the state has plans, training, and exercises, (2) mitigation in which a sustained effort is made to reduce risks to people and property, (3) response during which immediate action is taken to save lives and protect properties, and (4) recovery in which post-earthquake reconstruction becomes the central focus. In the case of the 2008 Wenchuan earthquake, the reactions of the central, provincial, and local governments were swift in the response and recovery stages. The governments of China and Japan were both unprepared in the coming of the Wenchuan earthquake and the 2011 tsunami respectively. However, mitigation measures have been considered and taken since the recovery stage, like careful rural land planning and designs taking into the ecological environment in China's Sichuan province and better Japanese coastal settlement planning in the wake of the unpredictable height of tsunami waves.

Col argued that the lessons learned from the Hurricane Katrina and the 1976 Tangshan earthquake were, firstly, the need for local governments to act decisively and earlier in preparation for any natural disasters, and secondly the support by higher-level governments in disaster preparation and mitigation efforts, and thirdly, the participation of citizens in all the phases of preparation and implementation of emergency management measures. Both China and Japan have learned from the lessons of the Wenchuan earthquake in 2008 and the earthquake tsunami in 2011. Similarly, the PRC government has since the Wenchuan tragedy been learning quickly, including the formulation and implementation of better public policy for the sake of reducing earthquake risks, such as the amendment of the Law on Earthquake Prevention and Disaster Management, especially the original Article 22 on the need for detailed planning so as to reduce earthquake risks. Moreover, building codes in Sichuan province are now strictly enforced. Other measures include the need for better land-use planning, the adoption of protective measures on the part of public utilities, large corporations, museums and educational institutions, and the enforcement of mitigation plans (Comerio, 2004: 403). In Sichuan province, medical clinics dealing with the earthquake victims also cooperated with and learned from the Hong Kong Red Cross on how to deal with their recovery and rehabilitation (*ATV News*, May 24, 2008). Policy learning has emerged since the outbreak of the Wenchuan earthquake and such learning is crucial for the PRC to better prepare for the occurrence of any natural disaster in the future.

Ozerdem and Barakat have stressed the importance of sustainable development policy, which should ideally be implemented before any natural disaster to address poverty issues (Ozerdem and Barakat, 2000: 427). They also point to the need for sustainable development after earthquakes so that reconstruction work can target at those who need the state's assistance and rescue.

The socialist regime of the PRC has paid considerable attention on the alleviation of poverty, and as such, sustainable development was already implemented before the 2008 Wenchuan earthquake, although the income gap between the rich and the poor was widened in the process of economic modernization and development throughout the 1990s and 2000s.

Yet, if sustainable development embraces the care and concern of the government about minority groups, the Wenchuan earthquake did demonstrate the PRC's swift responses to the predicament of ethnic minorities. After the Wenchuan earthquake, special care was paid to deal with the Qiang's cultural heritage. The message was clear: in order to maintain regime legitimacy, the PRC's Party-state has to constantly care about the interests of ethnic minorities.

If the occurrence of earthquakes can be seen as the roar of the nature, they also test the responses of the political dissidents, and activists who challenge the legitimacy of the regime in power. The Wenchuan earthquake did open a window of opportunity for the Dalai Lama and his supporters to carve a moderate path or an intermediary way between the pro-Tibetan independence activists and the relatively hard-line PRC government. Yet, this strategy adopted by the Dalai Lama was futile as the PRC government believed that the Tibet Autonomous Region (TAR) was governed with a high degree of CCP-state legitimacy, and that any "one country, two systems" model as propounded by the Dalai Lama was tantamount to making the TAR separate from the PRC sphere of influence. In the minds of the Beijing leaders, although the Dalai Lama and his supporters were and are more moderate than the radical independence activists, the Tibet issue has been actually utilized by the Tibetan exile government in collaboration with some foreign states to split the PRC territorially. As such, Beijing's hard-line attitude toward the question of Tibet's future is understandable. Still, even though some Tibetans strive for independence, the PRC administration has been continuing to hold talks with the moderate Dalai Lama and his representatives so that dialogue and communication can hopefully minimize differences and achieve the Party-state's objective of conducting united front tactics, namely wooing potential enemies into the orbit of political friends. The fact that the Dalai Lama appealed to the radical independence activists to mourn the dead people immediately after the Wenchuan earthquake was a sign of political reconciliation. The PRC government, on the other hand, also hoped that the Dalai Lama could be the intermediary appealing to the pro-Tibet independence activists inside the TAR for calmness and serenity. Hence, the politics of the Wenchuan earthquake included not only the roles of central, provincial, and local governments in their responses but also the reactions from the political activists, dissidents, and "separatists."

Internally, the PRC regime at the provincial and local levels remained relatively intolerant of political dissidents critical of the "tofu dreg" projects, a term coined by Tan Zuoren who was arrested in 2009 in Chengdu for his activities of instigating public dissent through the Internet. The PRC authorities arrested him on the charge of defaming the CCP and raising the issue of the 1989 Tiananmen incident in public. He was imprisoned for five years until March 2014 (Hong Kong Cable TV News, March 26, 2014). It is crystal clear that the PRC regime remains a paternalistic one, like a father who is determined to deal with his family affair but is relatively intolerant of any dissenting voices from his sons and daughters. If paternalism is a hallmark of Chinese politics, it also shapes the way in which the Party-state deals with natural disasters, such as earthquakes.

Finally, some social scientists have focused on the preparedness of governments in anticipation of, and in response to, earthquake disasters. Michael Hooper studied the 1994 Northridge earthquake in California and found that residents had been educated about the possible occurrence of any earthquake (Hooper, 1999). Richard Vedder also utilized the case of the 2005 Hurricane Katrina to highlight the significance of the need to improve disaster planning (Vedder, 2006). Strictly speaking, the ways in which the PRC government responded to the Wenchuan earthquake had tremendous educative functions. Citizens were not only educated about how they could deal with natural disasters, but the governments at all levels were also learning how the rescue, relief, and reconstruction work should be implemented. Comprehensive mobilization of citizens, groups, media, PLA, police, and governments at various levels turned out to be the most important educative campaign in the entire process of coping with the consequences of the Wenchuan earthquake. These educative aspects cannot be swept into the carpet as policy learning, disaster management experiences, and post-earthquake reflections and research can contribute to, in the words of Hooper and Vedder, the improvement in governmental preparedness and disaster planning in the long term.

In short, the study of the 2008 Wenchuan earthquake is important academically and practically. From the perspective of political science, the ways in which the governments at the central, provincial, and local levels responded to the Wenchuan earthquake with prompt action and resilience represented the crucial factor shaping the consolidation of the PRC's regime legitimacy. Most impressively, the mobilization of provinces, cities, mass media, NGOs, volunteer groups, individual citizens, the PLA, the police, and pro-establishment transmission belts such as women and labor unions became the effective tools through which the Party-state conducted the rescue, relief, and reconstruction work swiftly and efficiently. The Party-state combined a managerial approach that emphasized efficiency and effectiveness with a political approach of public mobilization. Legally, the PRC's responses to the Wenchuan earthquake were in conformity with the content of the Law on Earthquake Prevention and Disaster Management, particularly chapters 4 and 5 dealing with emergency response and relief and reconstruction respectively. The focus of chapter 4 on the command and institutional responses from the State Council and government departments at the central, provincial, and local levels was all strictly followed after the Wenchuan earthquake. Similarly, the emphasis of chapter 5 on organization, communication, and coordination of all departments responsible for relief and reconstruction work was prominent in the days, weeks, and months after the Wenchuan earthquake. Hence, it can be said that the Chinese amalgamated the managerial, political, and legal approaches in their responses to the Sichuan earthquake in 2008, thus consolidating the regime legitimacy overall.

The implications of the PRC ways of responding to the Wenchuan earthquake are significant. First and foremost, although the PRC seismological authorities failed to anticipate the occurrence of the Sichuan earthquake before May 12, 2008, the swift reactions and mobilization capacity of the Party-state after the earthquake served as the mitigation factor that minimized the damages caused by the natural disaster on the people's livelihood, society, and economy. Second, the mobilization capacity of the Party-state did have a positive impact on the regime legitimacy, unlike the 2010 Haiti earthquake that required the immediate external assistance from the American military and the 2011 tsunami that produced a nuclear power plants crisis necessitating the technological and military backup from the United States. To a large extent, the PRC government handled the consequences of the Wenchuan earthquake through its own internal efforts, although foreign aid and logistical supplies were also warmly embraced and welcome. Third, the PRC's mobilization strategy after the Wenchuan earthquake had implications for socialist regimes like Myanmar, which tended to lack mobilization capacity after the cyclone's attack in May 2008. Building up the regime's mobilization capability is perhaps a tactic that can be employed by other socialist regimes in dealing with natural disasters, for the Chinese experiences provided an important policy learning experience.

Trevor Jones and Tim Newburn have arrived at a number of insightful conclusions on the concept, mechanisms, and limitations of policy learning and transfer. In general, policy transfer occurs primarily at the levels of "policy ideas, symbols and labels," and therefore "soft" transfer in terms of initial ideas and principles is more common than "hard" transfer in the aspects of adopting the "actual nuts and bolts of policy" (Jones and Newburn, 2006: 143–145). The mechanisms of policy transfer, according to Jones and Newburn, include government officials and politicians. The processes of policy transfer include elite networking, conferences, seminars, exchange programs, study visits, think tanks, and even international organizations (Jones and Newburn, 2006: 149–151). Jones and Newburn conclude that "while it seems clear that policy ideas and innovations travel both within and between polities, they undergo a range of reshaping once they disembark"

(Jones and Newburn, 2006: 158). The limits to policy transfer, according to Newburn and Jones, embrace "political, legal and institutional contexts" that shape the extent to which policy ideas, content, and instruments can be transplanted from one place to another (Jones and Newburn, 2006: 154).

If we apply the concepts of Jones and Newburn to explore the applicability of the Chinese ways of responding to the Wenchuan earthquake to other settings, including socialist and non-socialist regimes, several observations can be made. First and foremost, both "soft" transfer of policy ideas and "hard" transfer in the dimension of adopting the "actual nuts and bolts of policy" can be learned and experimented in another country. For instance, as argued above, the socialist regime of Myanmar could have learned much from the mobilization capacity and strategy of the PRC government in dealing with natural disasters. The idea that all citizens and groups should be mobilized and involved in the rescue, relief, and reconstruction work can indeed be transplanted into another setting like Myanmar. Second, better public policy in terms of urban planning, rural landscape usage, educating the public about how to prevent the harmful impacts of natural disasters like earthquakes, and psychological counseling and support for both citizens and government personnel can all be transplanted into another setting. If natural disasters can cut across national boundaries, the policy transfer experiences dealing with natural disasters are arguably the same. Indeed, not everything can be copied from the PRC experiences, as the political, legal, and institutional contexts may be quite different among countries. The Japanese legal, political, and institutional responses to natural disasters are bound to be different from the Chinese. However, this does not mean that scientists, government officials, citizens, and groups cannot learn from each other in their experiences in coping with natural disasters. Better preparation, planning, responses, and reconstruction work can be seen as the policy ideas and principles that are shared by countries in managing natural disasters. Hence, future research agenda and directions may focus on the technicalities involving the "actual nuts and bolts of the policy" that can be mutually shared by government officials, scientists, seismological officials, citizens, and groups in their joint efforts at preventing and managing the occurrence of earthquakes in the coming years.

Appendix 1

The PRC Law on Earthquake Prevention and Disaster Management

Translated by the author from the law as published in www.gove.cn, September 27, 2005.

Approved by the Standing Committee of the National People's Congress on December 29, 1997 and promulgated by the PRC President in Promulgation Number 94, with the effect of implementation from March 1, 1997.

CONTENT

Chapter 1: Principles

Article 1: This law is formulated in order to prevent and reduce earthquake disaster, protect the lives of citizens and their properties' safety, and to safeguard the smooth operation of socialist construction.

Article 2: Those activities which can detect earthquake, prevent earthquake disaster, respond to earthquake, and conduct post-earthquake reconstruction work are suitable for this law.

Article 3: Earthquake reduction work has to focus on prevention and adopts the principle of combining prevention with relief.

Article 4: Earthquake reduction work should be integrated into the national economic and social development plans.

Article 5: The nation encourages and supports scientific and technical research on earthquake reduction and it promotes the fruits of scientific research to enhance the level of earthquake reduction and prevention.

Article 6: The governments at all levels should take the leadership of earthquake prevention and reduction, organize other departments concerned to adopt appropriate measures to conduct earthquake prevention and reduction work.

Article 7: Under the leadership of the State Council, the seismological department, economic department, construction department, civil affairs departments, and others should have clear division of labor and coordinate with each other to conduct better earthquake prevention and reduction work. The governments above the county level should take the leadership of achieving division of labor and coordination work so as to conduct better earthquake prevention and reduction work.

Article 8: Any unit or individual can have the obligation according to the law to participate in activities to prevent and reduce earthquake. The People's Liberation Army and the People's Armed Police should implement the responsibility conferred by the nation of dealing with earthquake prevention and reduction.

Chapter 2: Earthquake Detection and Report

Article 9: The nation consolidates earthquake detection and report work, encourages and supports scientific report of earthquake detection so as to increase the standard of earthquake prediction.

Article 10: The State Council and the seismological departments concerned formulate the plan of national earthquake detection and report and implement it. The people's governments at the provincial, special autonomous regions, and directly administered cities are responsible for managing the departments that detect and report on earthquake and implementing the earthquake detection plans.

Article 11: The State Council's seismological departments give concrete opinions on earthquake's epicenter and prevention work to the State Council which then approves their plans. The seismological departments above the county level are responsible for enhancing earthquake detection work, formulating short-term earthquake prevention plans, and establishing earthquake tracing system so as to improve the report system of earthquake occurrence.

Article 12: The State Council and seismological departments above the county level should inspect, transfer, analyze, and arrange the data with regard to earthquake detection and occurrence.

Article 13: The nation implements the management of centralized planning, categorization, and classifications of setting up earthquake detection stations. The earthquake detection and supervision stations are established by the nation's basic station networks and the related networks at the provincial, county, and city levels. The central and local financial administrations are responsible for the investment, finance, and management of these stations. The investment and management of these stations are under the leadership and guidance of the seismological departments of the people's governments above the county level.

Article 14: The nation supervises the earthquake detection facilities and environment, and no unit or individual can endanger these facilities and environment. The earthquake detection environment is designated as protective areas in accordance with the requirements of having no influence and no disturbance on the efficiency of the work of the earthquake detection and supervision stations. The earthquake detection and supervision facilities refer to those instruments designated by the State Council's seismological and administrative departments.

Article 15: The newly expanded, constructed, and reconstructed work should avoid endangering the earthquake detection and supervision facilities and environment. If it is impossible to avoid endangering these facilities and environment, the unit concerned has to seek the approval of the State Council's seismological department and the responsible departments above the county level and to follow the State Council's subsequent measures.

Article 16: The nation has to implement a centralized report system on short-term earthquake prediction and situation, and it has to be announced by the people's governments at the provincial, special autonomous regions, and directly administrated cities in accordance with the regulations of the State Council. Any unit or seismological professionals having opinions on the short-term earthquake predictions and detections should not be disseminated to the society at their own will without prior reports to the State Council's seismological authorities and the people's governments above the county level.

Chapter 3: Earthquake Prevention

Article 17: Any newly established, expanded, and reconstructed work has to conform to the standard of preventing and resisting earthquake. Any construction work beyond Article 3 of this law has to conform to the standard of earthquake resistance so as to implement earthquake prevention measures. If any important construction work may incur serious side effects on natural disasters, it has to be assessed in terms of safety during earthquake so as to implement preventive measures in accordance with the requirement of earthquake prevention. The so-called important construction work refers to projects with significant value or influence on the society. The natural disasters that have side effects from construction work refer to the reservoirs, large dams, oil and gas stations that may produce flooding, fire accidents, explosion, poisonous, or erosive chemical leakages. The construction work near nuclear power stations or facilities that may produce radioactive pollution and disasters has to be assessed in terms of earthquake safety, and it should implement earthquake resistance measures.

Article 18: The nation's seismological department is responsible for drawing up the blueprint for earthquake-prone zones so as to carry out earthquake safety work based on the assessment results. The State Council's construction department is responsible for the blueprint for housing construction and the related design that can resist earthquake, except for the work as prescribed in this law's Article 3. The State Council's railroad, transport, civilian aviation, water supply, and other professional departments are responsible for the blueprint and designs of railways, highways, ports, piers, airports, and waterworks to resist earthquake.

Article 19: Any construction work has to follow the design and requirements of resisting earthquake.

Article 20: For those construction work and buildings that have not implemented earthquake resistance measures, they should in accordance with the nation's requirements adopt the necessary earthquake prevention remedial and consolidation measures to encompass the following areas: (1) large-scale construction, infrastructure and buildings; (2) those buildings and infrastructure work that may have side effects on natural disasters; (3) those buildings and infrastructure work with great cultural and historical value and meanings; and (4) those buildings and infrastructure in regions under the close supervision of earthquake prevention and detection.

Article 21: The people's governments concerned should adopt effective preventive measures to deal with natural disasters which may be impacted by earthquake, such as fire hazards, flooding, landslide, radioactive pollution, and infectious diseases.

Article 22: In accordance with the earthquake prediction and actual situation, the State Council's seismological department and the people's governments above the county level should cooperate in preparing earthquake reduction and prevention plans, which are then approved and implemented by the people's governments. The revision of any earthquake prevention and reduction blueprint should be reported and approved by the departments and authorities concerned.

Article 23: The people's governments at all levels should carry out propaganda and education on how to prevent earthquake and reduce natural disasters so as to increase the awareness of citizens and enhance their self-aid and self-rescue ability as well as to consolidate the training of the professionals concerned, thus elevating the capability of rescue and relief operations.

Article 24: The people's governments above the county level can arrange proper anti-earthquake relief and logistical supplies in accordance with the practical needs and current financial budgetary circumstances.

Article 25: The nation encourages units and individuals to participate in insurance against earthquake disasters.

Chapter 4: Earthquake Emergency Response

Article 26: The seismological department and other related agencies under the State Council formulate emergency response plans to deal with the disruptive impacts of earthquake, and such plans have to be reported and approved by the State Council. The State Council's agencies also submit the emergency response plan to the superior authorities for reporting and approval. Similarly, the departments at the people's governments in cities with over one million population above the county level have to follow the national emergency response plans to formulate corresponding emergency response blueprints that should also be submitted to both the people's government and the State Council for reporting and approval. Disruptive impacts of earthquake in this law refer to the earthquake disasters bringing about a large number of deaths and tremendous loss of properties.

Article 27: The nation encourages and supports emergency response work, including rescue technique, facilities, and studies. The people's governments above the county level have the responsibility of formulating emergency response plans, and the storage of rescue facilities, and the training in the usage of these rescue facilities.

Article 28: The emergency response plan dealing with disruptive earthquake includes the following content: (1) the organization and responsibilities of emergence response institutions; (2) the safeguard of emergency response telecommunications; (3) the preparation of the organization, capital, and logistical supplies of rescue teams and workers; (4) the preparation of emergency response and rescue facilities; (5) the preparation of disasters assessment; and (6) the blueprint for emergency response action.

Article 29: After the earthquake detection and prediction reports are announced, the people's governments in provinces, autonomous regions, and directly administered cities need to promulgate the inception of the pre-earthquake emergency period and to conduct preparatory work for rescue operation and the mobilization of social forces to deal with the coming earthquake.

Article 30: After the serious damage and loss created by earthquake, the State Council should set up a command headquarters to counter the earthquake and to organize the departments concerned to implement the emergency response plans. The offices of the leading institutions under the State Council should be set up inside its seismological administration department. Similarly, the people's governments above the county level should also set up commanding headquarters and organize the departments concerned to implement the emergency response plans after the disruptive earthquake. The disruptive earthquake in this law refers to the death tolls, the loss of properties, the loss of self-renewal capability of disaster regions, and the need for the nation to take corresponding actions to tackle the earthquake disasters.

Article 31: The people's governments at all levels should timely report the earthquake situation and disaster circumstances to the people's governments at the upper level. The people's governments in provinces, autonomous regions, and directly administered cities should also follow the State Council's regulations to report to the society about the earthquake situation and disaster circumstances. The departments in the provinces, autonomous regions, and directly administered cities should also investigate the disaster situation, make assessments, and submit reports to the people's government at the same level.

Article 32: After the disruptive earthquake's occurrence, in order to maintain social order, the State Council or the people's governments in the affected provinces, autonomous regions, and directly administered cities can implement the following emergency measures: (1) transport control; (2) the centralized distribution and allocation of all basic necessities such as food and medicine; (3) the temporary requisition of houses, transport vehicles, and telecommunication facilities; and (4) the adoption of other necessary emergency response measures.

Chapter 5: Earthquake Relief and Reconstruction

Article 33: After the occurrence of disruptive earthquake, the people's governments in affected regions at all levels should organize the local forces, rescue teams, and self-aid and self-rescue personnel, while the people's governments in non-affected regions should also organize and mobilize social forces so as to provide assistance and aid in accordance with the earthquake situation and disaster circumstances. After the occurrence of disruptive earthquake, the State Council should also provide rescue assistance and urge the economic departments concerned to coordinate with other departments to arrange the transport of disaster relief and logistical supplies.

Article 34: The people's governments above the county level in earthquake-affected regions should also organize the health, medical, and other related departments and units and conduct proper rescue and health work in prevention against infectious disease.

Article 35: The people's governments above the county level in earthquake-affected regions should organize the department of civil affairs and other related departments to rapidly set up refuge centers and provision points for relief and logistical supplies so that the livelihood, resettlement, and reallocation of earthquake victims can be properly arranged.

Article 36: The people's governments above the county level in earthquake-affected regions should organize the departments of transport, mail delivery, electricity, and construction to adopt measures that could quickly resume the damaged transport, telecommunication, water supplies, drainage system, electricity supplies, gas supplies, and the transport of oil and gases, and to take emergency preventive measures against any other natural disasters caused by the earthquake.

Article 37: The people's governments above the county level in earthquake-affected regions should organize the public security agencies and other related departments to strengthen the management of law and order, the protection of safety, the prevention and combat against other types of criminal activities, and the safeguard of social order.

Article 38: For the temporarily requisitioned housing, transport vehicles and telecommunication facilities, they have to be timely returned. Compensation or proper arrangement has to be made to those whose requisitioned properties are damaged in accordance with the regulations of the State Council.

Article 39: In the process of post-earthquake relief operation, any unit or individual has to obey the law, social morality, orders, and to protect social order voluntarily.

Article 40: Any unit or individual cannot terminate or embezzle the capital and logistics in the earthquake relief operations. The people's government at each level has to strengthen the audit work and supervision of the usage of capital and logistics for the relief operations.

Article 41: The people's governments above the county level in earthquake-affected regions should coordinate the planning and arrange reconstruction work in accordance with the earthquake situation and the requirements of countering the earthquake.

Article 42: The nation has to protect the typical earthquake zones and sites, which should be integrated into the reconstruction plans.

Chapter 6: Legal Liability

Article 43: Those who violate this law in the following way would be terminated by the State Council's seismological department or the people's governments above the county level, with serious circumstances having a penalty of 5,000 yuan to 100,000 yuan or civilian responsibility or criminal liability: (1) incurring damage illegally to newly constructed, expanded, or reconstructed work or to earthquake detection facilities and observation environment; and (2) incurring damage to the typical earthquake sites and legacies.

Article 44: A penalty of 10,000 yuan to 100,000 yuan would be imposed by the State Council's seismological department or the people's governments above the county level on those who violate this law's Article 17, and who would have to correct any action on the part of the construction unit that ignores earthquake safety assessment or neglects the anti-earthquake design based on earthquake safety assessment.

Article 45: A penalty of 10,000 yuan to 100,000 yuan will be imposed by the people's governments above the county level or by the construction agencies and their associated professional departments on those who ignore the anti-earthquake designs in accordance with the earthquake prevention blueprint, and those who implement construction work ignoring the anti-earthquake design.

Article 46: Those who terminate and squeeze the relief capital and logistical supplies would have criminal liability or would face administrative penalties.

Article 47: Criminal liability would be pursued or administrative penalties would be imposed on those national workers who abuse their powers, who are administratively negligent, and who commit offense for the sake of achieving private gains.

Chapter 7: Supplementary Principle

Article 48: This law has been implemented since March 1, 1998.

Bibliography

NEWS AGENCIES, NEWSPAPERS AND MAGAZINES

Apple Daily (Hong Kong Chinese newspaper).
ATV News (Asia Television News, Hong Kong).
Beijing Review (China's weekly magazine).
CCTV (Beijing's official broadcasting television news).
Cheng Pao (Hong Kong Chinese newspaper).
China Daily (mainland Chinese newspaper).
Globe and Mail (Canadian English newspaper).
Hong Kong Cable TV News (Hong Kong Cable Television).
Hong Kong Commercial Daily (Hong Kong Chinese newspaper).
Hong Kong Daily News (Hong Kong Chinese newspaper).
Hong Kong Economic Journal (Hong Kong Chinese newspaper).
Japan Times (Japan's English newspaper).
Ming Pao (Hong Kong Chinese newspaper).
People's Daily (China's newspaper).
Reuters (International news agency).
Sing Tao Daily (Hong Kong Chinese newspaper).
Ta Kung Pao (Hong Kong Chinese newspaper).
The Sun (Hong Kong Chinese Newspaper).
The Toronto Sun (Canada's Toronto English newspaper).
Today Daily News (Toronto's version of Hong Kong-based *The Sun* newspaper).
Toronto Star (Canadian Toronto English newspaper).
Wen Wei Po (Hong Kong Chinese newspaper).
World Journal Monthly (Toronto Chinese weekly magazine).
Xinhua (China's news agency).
Zhonghua Tangshuo (*China Exploration*, a Toronto Chinese monthly magazine based on the *Yazhou Zhoukan* or *Asiaweek*).

CHINESE LITERATURE

Cao, Jun, et al. (2008). *Globalization and National Security in China*. Beijing: Social Sciences Academic Press.

Chen, Qiang (2008). "The Progress and Principles of Emergency Management in the Wenchuan Earthquake,"*Fujian Forum: Humanities and Social Sciences Edition*, Special Issue, pp. 35–36.

Chen, Sheng, and Meng, Qingguo (May 2010). "A Study on the Evaluation of Local Government Capacity in Abnormity: Reconstruction after Wenchuan Earthquake for Example," *Academics*, vol. 5, no. 144, pp. 36–45.

Gong, Zheng, and Wu Sibian, eds. (2008). *The Truth Behind the Sichuan Earthquake*. Hong Kong: Global Enterprises.

Lin, Mingang, and Zhan, Jianhua (2010). "NGOs Participation in Disaster Relief and Relevant Management: The Cases of Wenchuan Earthquake and the September 21 Earthquake in Taiwan," *Chinese Public Administration*, vol. 3, no. 297, pp. 98–103.

Qian, Gang (2010). *The Great Earthquake of Tangshan*. Hong Kong: Chung Hwa Book.

Wang, Xiangping (2005). "Connecting Emotions with the People: Some Details of Zhou Enlai During the Xingtai Earthquake's Rescue Process," *Literature of the Chinese Communist Party*, no. 4, pp. 118–119.

Wu, Jianxun, and Xu, Xiaodi (2010). "On the Relationship between Non-Profit Organizations and Governments in the Public Crisis Management in China: A Study Based on the Analysis of 33 NPOs' Performances in Wenchuan Earthquake," *Theory, Methodology and Cases*, vol. 10, no. 57, pp. 347–362.

Zhang, Qiang; Lu, Qibin; and Zhang, Xiulan (2011). "An Outlook on Chinese Model of Emergency Management Based upon the Review of Wenchuan Earthquake Response: An Interactive Perspective of Strong State and Strong Society," *Chinese Public Administration*, vol. 5, no., 311, pp. 50–56.

Zhao, Yongsheng (2008). "The Days and Nights of Zhou Enlai after the Xingtai Earthquake," *The Chinese Communist Party's History*, August, pp. 4–8.

Zhou, Jiading (1997). "In Memory of Premier Zhou Enlai's Two Visits to the Regions Affected by the Xingtai Earthquake," *Literature of the Chinese Communist Party*, no. 5, pp. 78–81.

ENGLISH LITERATURE AND SOURCES

Arbour, Mary Catherine; Murray, Kara; Arriet, Felipe; Moraga, Cecilia; and Vega, Miguel Cordero (2011). "Lessons from the Chilean Earthquake: How a Human Rights Framework Facilitates Disaster Response," *Health and Human Rights*, vol. 13, no. 1 (July 2011), pp. 62–73.

Bapat, Arun (2010). "Disaster Management Plans in View of Recent Earthquakes," *Current Science*, vol. 98, no. 10, pp. 1287–1288.

Carter, Liz (2013). "Does Anyone Trust the Red Cross?," May 2, 2013, in http://www.theatlantic.com/china/archive/2013/05/does-anyone-trust-the-chinese-red-cross/275508/, access date: June 21, 2014.

Chase, Michael S. (2012). "Army Day Coverage Stresses PLA's Contributions and Party Control," *China Brief*, vol. 12, no. 16, in http://www.jamestown.org/single/?tx_ttnews%5Btt_news%5D=39780&no_cache=1#.U6WozZqwoqQ, access date: June 28, 2014.

Chen, Qi-Fu; Mi, Hongliang; and Huang, Jing (2005). "A Simplified Approach to Earthquake Risk in Mainland China," *Pure and Allied Geophysics*, no. 162, pp. 1255–1269.

Choate, Allen (2011). "In Face of Disaster, Japanese Citizens and Government Pull from Lessons Learned," in http://asiafoundation.org/in-asia/2011/03/16/in-face-of-disaster-japanese-citizens-and-government-pull-from-lessons-learned/, access date: March 23, 2014.

Col, Jeanne-Marie (December 2007), "Managing Disasters: The Role of Local Government," *Public Administration Review*, Special Issue, pp. 114–124.

Col, Jeanne-Marie, and Chu, Jean J. (2001), "Integrating Public Administration, Science, and Community Action," in Ali Farazmand, ed., *Handbook of Comparative and Development Public Administration*. New York: Marcel Dekker.

Comerio, Mary C. (2004), "Public Policy for Reducing Earthquake Risks: a US Perspective," *Building Research & Information*, vol. 32, no. 5, September/October, pp. 403–413.

Constitutional and Mainland Affairs Bureau, and Development Bureau (December 2011). "For Information: Legislative Council Panel on Development Progress Report on the HKSAR's Work in Support of Reconstruction in the Sichuan Earthquake Stricken Areas," Legislative Council document, CB(1)575/11–12(01).

Constitutional and Mainland Affairs Bureau (July 2008). "Legislative Council Brief: HKSAR to Support Reconstruction in the Sichuan Earthquake Stricken Areas," in http://www.legco.gov.hk/yr07-08/english/fc/fc/papers/fc0714fc-112-e.pdf, access date: June 21, 2014.

Esposito, Anthony and O'Brien, Rosalba (2014). "Chile Hit By Powerful Quake Aftershock, No Major Damage," *Reuters*, April 2.

Feickert, Andrew and Chanlett-Avery, Emma (2011). *Japan 2011 Earthquake: U.S. Department of Defense (DOD) Response*. Washington: Congressional Research Service Report for Congress, in www.crs.gov.

Government of Japan (2011). *Lessons Learnt from the Great East Japan Earthquake: Hazard Information and Damage Scenarios to Inform Effective Countermeasures to Extreme Events*, September 11, in http://www.gfdrr.org/sites/gfdrr.org/files/Chapter_11-Japan-Lessons_Learned_from_the_Great_East_Japan_Earthquake.pdf, access date: March 20, 2014.

Heath, Robert (1995). "The Kobe Earthquake: Some Realities of Strategic Mismanagement of Crises and Disasters," *Disaster Prevention and Management*, vol. 4, no. 5, pp. 11–24.

Henao, Luis Andres (2014). "Chile Earthquake 2014: Tsunami Warnings Prompt Evacuation Orders," *Associated Press*, April 2.

Hinrichs, Richard; Jones, Lucy; Stanley, Ellis M.; and Kleiner, Michael (2011). *Report on the 2010 Chilean Earthquake and Tsunami Response*, United States Department of Interior and United States Geological Survey, Open File Report 2011–1053, in http://pubs.usgs.gov/of/2011/1053/of2011-1053.pdf, access date: June 28, 2014.

Hooper, Michael K. (1999). "Disaster preparedness: an analysis of public safety agency and community preparedness during the Northridge earthquake," *International Journal of Public Administration*, vol. 22, no. 5, pp. 679–710.

Hosseini, Mahmood, and Izadkhah, Yasamin O. (2006), "Earthquake Disaster Risk Management Planning in Schools," *Disaster Prevention and Management*, vol. 15, no. 4, pp. 649–661.

Jinde Charities (2010). "Earthquake Relief in Yushu, Qinghai Province, China," in http://eng.jinde.org/uploadfile/2012/0524/20120524022750143.pdf, access date: June 22, 2014.

Jones, Trevor, and Newburn, Tim (2006). *Policy Transfer and Criminal Justice*. Buckingham: Open University Press, 2006.

Kaufmann, Daniel, and Penciakova, Veronika (2011). "Japan's Triple Disaster: Governance and the Earthquake, Tsunami and Nuclear Crises," http://www.brookings.edu/research/opinions/2011/03/16-japan-disaster-kaufmann, access date: March 24, 2014.

Kim, Pan Suk, and Lee, Jae Eun (2001). "Emergency Management in Korea: Mourning over Tragic Deaths," in Ali Farazmand, ed., *Handbook of Comparative and Development Public Administration*. New York: Marcel Dekker.

Morelli, Ugo (2004). "Preventing Earthquake Disasters—The Grand Challenge in Earthquake Engineering—A Research Agenda for the Network for Earthquake Engineering Simulation," *Emergency Management*, vol. 1, no. 4, pp. 1–4.

Mulvenon, James (2010). "Party-Military Coordination of the Yushu Earthquake Response," *China Leadership Monitor*, June 28, in http://media.hoover.org/sites/default/files/documents/CLM33JM.pdf, access date: June 22, 2014.

NDIS China Security Report 2012 (2012). Japan: National Institute for Defense Studies.

Nakamura, Akira (2000). "The Need and Development of Crisis Management in Japan's Public Administration: Lessons from the Kobe Earthquake," *Journal of Contingencies and Crisis Management*, vol. 8, no. 1, pp. 23–29.

Nateghi-A, Fariborz (2000). "Existing and Proposed Disaster Management Organization for Iran," *Disaster Prevention and Management*, vol. 9, no. 3, pp. 200–204.

Nateghi-A, Fariborz, and Izadkhah, Yasamin O. (2004). "Earthquake Disaster Management Planning in Health Care Facilities," *Disaster Prevention and Management*, vol. 13, no. 2, pp. 130–135.

Lo, Sonny Shiu-Hing (August 2007). "Infectious Disease in Asia: Policy Implications for Canada," Research Report of the Asia Pacific Foundation of Canada, in http://www.asiapacific.ca/analysis/pubs/listing.cfm?ID_Publication=617, access date: April 28, 2014.

Nakamura, Akira, and Kikuchi, Masao (2011). "What We Know, and What We Have Not Yet Learned: Triple Disasters and the Fukushima Nuclear Fiasco in Japan," *Public Administration Review*, November/December, vol. 71, no. 6, pp. 893–899.

Olson, David L., and Wu, Desheng Dash (2010). "Earthquakes and Risk Management in China," *Human and Ecological Risk Management: An International Journal*, vol. 16, no. 3, pp. 478–493.

Ozerdem, Alpaslan, and Barakat, Sultan (2000). "After the Marmara Earthquake: Lessons for Avoiding Short Cuts to Disasters," *Third World Quarterly*, vol. 21, no. 3, pp. 425–439.

Pomonis, Antonios (2006). "The Mount Parnitha (Athens) Earthquake of September 7, 1999: A Disaster Management Perspective," *Natural Hazards*, no. 27, pp. 171–199.

"PLA military regions" (2002), in http://news.xinhuanet.com/ziliao/2002-01/25/content_254347.htm, access date: June 21, 2014.

Saavedra, Jose Luis (2014). "Earthquake Rocks Chile," *Reuters*, April 1.

Saich, Tony (2006). "Is SARS China's Chernobyl or Much Ado About Nothing," in Arthur Kleinman and James L. Watson, eds., *SARS in China: Prelude to Pandemic?* Stanford: Stanford University Press, pp. 71–104.

Saito, Taiki (2007). "Disaster Management of Local Government in Japan," in http://www.oecd.org/dataoecd/55/60/37377837.pdf, access date: October 12, 2011.

Sakamoto, Masaru (2001). "Crisis Management in Japan: Lessons from the Great Hanshin-Awaji Earthquake of 1995," in Ali Farazmand, ed., *Handbook of Comparative and Development Public Administration*. New York: Marcel Dekker.

Sanchez, Ray (2014). "Experts: Strict Building Codes Saved Lives in Powerful Chile Earthquake," *CNN*, April 3.

Schneider, Florian, and Hwang, Yih-jye (2014). "The Sichuan Earthquake and the Heavenly Mandate: Legitimizing Chinese Rule through Disaster Discourse," *Journal of Contemporary China*, vol. 23, no. 88, pp. 636–656.

Song, Lin; Wang, Jianling; Yao, Shijie; and Chen, Jian (2012). "Market Reactions and Corporate Philanthropy: A Case Study of the Wenchuan Earthquake in China," *Journal of Contemporary China*, vol. 21, no. 74 (March), pp. 299–315.

Suganuma, Katsutoshi (2006). "Recent Trends in Earthquake Disaster Management in Japan," *Quarterly Review*, no. 9, April, pp. 91–105.

Tatsumi, Suki (November 2012). *Greater Eastern Japan Earthquake: "Lessons Learnt" for Japanese Defence Policy*. Washington: Stimson Center, in www.stimson.org.

Vedder, Richard (2006). "Katrina's Gift: A Wake-Up Call for Disaster Planning," *Information Security Journal: A Global Perspective*, vol. 15, no. 6, pp. 41–44.

Wang, Jian (2004). "Historical Earthquake Investigation and Research in China," *Annals of Geophysics*, vol. 47, numbers 2/3, April-June.

Wang, Wenhong, ed. (2013). *Chongjian Wenchuan* (*Reconstructing Wenchuan*). Hong Kong: Chung Hwa Book).

Wong, Raymond (2013), "Hong Kong Council's Anti-donation Speech," in http://www.youtube.com/watch?v=ro3xyYoY7lU, access date: June 22, 2014.

Yao, Chuanmei; Chen, Xunchui; and Yao, Lan (2009)."How China Responded to the May 2008 Earthquake During the Emergency and Rescue Period," *Journal of Public Health Policy*, vol. 30, no. 4, pp. 379–394.

Zare, M., and Afrouz, S. Ghaychi (2012). "Crisis Management of Tohoku: Japan Earthquake and Tsunami, 11 March 2011," *Iran Journal of Public Health*, vol. 41, no. 6, pp. 12–20.

Zhao, Hui (August 2009). *Chinese Government's Role in Crisis Management: Case Studies of Three Major Crises in Recent Years*. Unpublished M.A. Thesis in Strategic Public Relations, Faculty of Graduate School, University of Southern California.

Index

About the Author

Sonny Shiu-Hing Lo is professor and head in the Department of Social Sciences at the Hong Kong Institute of Education. He formerly worked at the University of Waterloo in Canada, the University of Hong Kong, the Hong Kong University of Science and Technology, Murdoch University in Australia, and the University of East Asia in Macau.

www.ingramcontent.com/pod-product-compliance
Ingram Content Group UK Ltd.
Pitfield, Milton Keynes, MK11 3LW, UK
UKHW012041150926
475560UK00001B/135

9 780739 139523